Nucleotide—One of the links in a DNA molecule. DNA contains four types abbreviated as A, T, G and C. Genetic information is stored in DNA through the specific order of nucleotides.

Protein—A chainlike molecule folded into a specific shape. Many proteins are responsible for directing the chemical reactions occurring in cells.

Ribosome—A workbench where proteins are made using information from DNA.

tRNA (transfer RNA)—A short chainlike molecule that acts as an adapter during the production of new proteins.

DNA polymerase—A protein responsible for making new DNA.

RNA polymerase—A protein responsible for making RNA, a chainlike molecule containing information copied directly from DNA.

DNA ligase—A protein molecule that splices broken DNA chains.

Repressor—A protein that binds to a specific spot on DNA and acts to prevent RNA synthesis from a specific gene.

E. coli—A type of microscopic organism (a bacterium) used to purify genes, to make trillions of copies of specific genes, and to produce large amounts of certain proteins.

Bacteriophage—A submicroscopic virus particle capable of infecting and killing bacteria like *E. coli*. Many bacteriophages are composed of DNA inside a protective protein coat. Bacteriophages serve as vehicles for inserting genes into *E. coli* cells.

Plasmids—Small circular DNA molecules that infect bacterial cells. Plasmids are used to insert genes into *E. coli*.

Understanding
DNA and Gene

CLONING

A Guide for
the Curious

KARL DRLICA

University of Rochester

JOHN WILEY & SONS

New York • Chichester • Brisbane • Toronto • Singapore

Library of Congress Cataloging in Publication Data:

Drlica, Karl.
   Understanding DNA and gene cloning.

   Includes index.
   1. Molecular cloning. 2. Genetic engineering.
I. Title. II. Title: Understanding D.N.A. and gene cloning.
[DNLM: 1. DNA, Recombinant.   2. Cloning, Molecular.
3. Genetic Intervention.   4. Molecular biology.
QH 442 D782u]
QH442.2.D75 1984   574.87′328   84-3518
ISBN 0-471-87942-8

Printed in the United States of America

10 9 8 7 6 5 4 3 2 1

To Ilene, for many years of support

PREFACE

An explosion of knowledge is shaking the science of biology, an explosion that will soon touch the life of each one of us. At its center is chemical information—information that our cells use, store, and pass on to subsequent generations. With this new knowledge comes the ability to manipulate chemical information, the ability to restructure the molecules that program living cells. Already this new technology is being used to solve problems in diverse areas such as waste disposal, synthesis of drugs, treatment of cancer, plant breeding, and diagnosis of human diseases. The new biology is also telling us how the chemicals in our bodies function; we may soon be programming ourselves and writing our own biological future. When this happens, each of us will be confronted with a new set of personal and political choices. Some of these difficult and controversial decisions are already upon us, and the choices will not get easier. Informed decisions require an understanding of molecular biology and recombinant DNA technology; this book is intended to provide that understanding.

Molecular biology is a science of complex ideas supported by test tube experiments with molecules. Consequently, the science has remained largely inaccessible to those without a knowledge of chemistry. I hope to change that situation—this book requires the reader to have little or no background in chemistry. Chemical processes and molecular structures are described by means of analogies using terms familiar to nonscientists. Technical terms have been kept to a minimum; where they must be introduced, they are accompanied by definitions. In addition, a glossary has been provided for easy reference; items in the glossary are in boldface type the first time they appear in the text.

It is also my intent to provide a sense of how informational molecules are manipulated experimentally. Integration of these details should help remove the mystery from gene cloning and expose the elegance and simplicity of the technology. I hope that this brief introduction to gene cloning will help you enjoy and appreciate the science of molecular biology for the art form that it is.

A number of people have helped me in this endeavor, and they deserve most of the credit for making this book readable. I especially thank Lynne Angerer, Betty Bonham, Tom Caraco, Cheryl Cicero, Lisa Dimitsopulos, Dianne Drlica, Karen Drlica, Rob Franco, Claire Gavin, Ed Goldstein, Brenda Griffing, George Hoch, Hiroko Holtfretter, Johannes Holtfretter, Lasse Lindahl, Stephen Manes, Bill Muchmore, Pat Pattison, Donna Riley, Peter Rowley, Ron Smith, Franklin Stahl, Todd Steck, Ilene Wagner, William Wasserman, Grace Wever, Bill Wishart, and Janice Zengel. I also acknowledge Alvin J. Clark and Henry Sobel for technical information and Ron Sapolsky for artistic insights used in early versions of the manuscript. Fred Corey and his staff at John Wiley & Sons provided excellent editorial assistance. The illustrations are the creative work of John Balbalis; where appropriate he has attempted to provide a sense of relative scale among the elements involved.

Karl Drlica

INTRODUCTION

In thinking about the course of human events, it has often occurred to me that they very much resemble the course of a river. A river meanders, gathers small streams, widens, deepens, and may even split into smaller rivers that go their separate ways. On occasion, rivers merge, a confluence that creates a mightier river. In the same sense, the extraordinary developments in genetic chemistry are part of an even more profound development in medical science, a change that is truly revolutionary. It is the confluence of the many discrete and previously unrelated medical science subjects into a single, unified discipline. Anatomy, physiology, biochemistry, microbiology, immunology, and genetics have now merged and are expressed in a common language of chemistry. By reducing structures and systems to molecular terms, all aspects of body form and function blend into a logical framework. Universities still maintain departmental lines to define administrative boundaries, but they are now meaningless in the pursuit of new knowledge.

The remarkable confluence of medical science first appeared in the genius of Louis Pasteur. More than any individual or school, he established medicine as a science and gave it the form we recognize today. Pasteur was trained as a chemist. His first exploit as a very young man was to show that two samples of tartaric acid of identical chemical composition differed physically because the molecules were mirror images of each other. Pasteur's "germ theory of disease" bore the stamp of his chemical background. He tried to reduce a problem of disease to elementary components. His experimental approach was to purify the causative agents to homogeneity and recreate the disease with the isolated pure form of the agent. From this he created and practiced the disciplines of microbiology and immunology. It might surprise many microbiologists and immunologists today to find that in 1911 the *Encyclopaedia Britannica* described Pasteur as a French *chemist*, the acknowledged head of the greatest *chemical* movement of his time.

Pasteur's scientific career had a flaw. Having established that the yeast cell is responsible for the conversion of sugar to alcohol, he tried to extract from the yeast cell the juices that would do the same thing. In this he failed and so concluded that nothing short of a living cell could possibly carry out this very complex chemical reaction. Pasteur's self-confidence, persuasiveness, and influence were so great that attempts by others to obtain alcoholic fermentation in a cell-free system were severely discouraged. And so, cellular vitalism became firmly rooted, and the advent of modern biochemistry was delayed for 30 years.

Only at the turn of this century did Eduard Büchner in Munich accidentally discover fermentation by disrupted yeast cells. In employing sugar as a preservative for yeast extracts, he observed a strange frothing. He had the insight to identify carbon dioxide as the gas and ethanol as the product of sugar degradation by the yeast juice. It was Pasteur's poor fortune that his extracts of Parisian yeast were deficient in sucrase, the enzyme that initiates the pathway of sugar metabolism. Luckily for Büchner, adequate amounts of the enzyme survived in his extracts from Munich yeast.

The reactions by which a yeast cell converts sugar to ethanol and carbon dioxide could then be isolated and analyzed in detail. In all, a dozen discrete, complex molecular rearrangements, condensations, and scissions are needed to achieve the fermentation of sugar to alcohol. Each of these reactions is catalyzed by an elaborate protein, an enzyme, designed to carry out the singular operation. The enzyme increases the rate of the reaction by a million- or trillionfold and gives it a unique direction among the many potential fates to which the molecule is susceptible.

These revelations of alcoholic fermentation in yeast provided the methods and confidence for the investigation of a comparable question. How does a muscle derive energy from sugar to do its work? When that mystery was unraveled, the plot and most of the characters in the muscle story incredibly proved to be the same as in yeast. There is, of course, one deviation. In muscle at the final stage, lactic acid is produced instead of alcohol and carbon dioxide.

Reconstitution in the test tube of the yeast and muscle pathways of sugar combustion to generate usable energy set the stage for a generation of enzyme hunters in the 1940s and 1950s. My own attempts at synthesizing DNA with enzymes in a test tube were regarded by some as audacious. Reconstitution of the metabolism of fats as well as carbohydrates may be one thing, but the enzymatic synthesis of genetically precise DNA, thousands of times larger, must be quite another. Yet all I have done is follow in the classical traditions of biochemists of this century. It always seemed to me that a biochemist with a devotion to enzymes could, with sufficient effort, reconstitute in the test tube any metabolic event as well as the cell does it. In fact, the biochemist, freed from the cellular restraints of the concentrations of enzymes, substrates, ions, and metals, and with the license to introduce reagents that retard or drive a reaction, should do it even better.

As the disciplines of genetics, microbiology, and physiology reached more and more for chemical explanations, they began to coalesce with the biochemistry of the enzyme hunters. From this coalescence came molecular biology and genetic engineering. Narrowing our focus to the molecular biology of DNA, I would cite several diverse origins. One origin is in medical science. In 1944 Oswald Avery, in his lifelong and relentless search for control of pneumococcal pneumonia, became the first to show that DNA is the molecule in which genetic information is stored. A second origin of molecular biology is in microbial genetics. In the late 1940s and early 1950s microbiologists, some of them renegade physicists, chose the biology of the small bacterial viruses, the bacteriophages, to elucidate the functions of the major biomolecules: DNA, RNA, and proteins. At about the same time a third origin of molecular biology arose as the structural chemistry of these biomolecules became highly refined. Analysis of the X-ray diffraction patterns of proteins revealed their three-dimensional structures; the DNA patterns gave us the double helix and a major insight into its replication and function. A fourth origin of molecular biology is in biochemistry, the enzymology, analysis, and synthesis of nucleic acids. The biochemist provided access to the nucleases that cut and dissassemble DNA into its genes and constituent building blocks, the polymerases that reassemble them, and the ligases that link DNA chains into genes and the genes into chromosomes; these are the reagents that have made genetic engineering possible. In the cell, these enzyme actions replicate, repair, and rearrange the genes and chromosomes.

Molecular biologists practice chemistry without calling it such. They identify and isolate genes from huge chromosomes, often only one part in millions or billions, and then they amplify that part by even larger magnitudes using microbial cloning procedures. They map human chromosomes, analyze their composition, isolate their components, redesign their genetic arrangement, and produce them in bacterial factories on a massive industrial scale. New species are created at will. Not even the boldest among us dreamed of this chemistry 10 years ago. I generally underestimated how permissive *E. coli* would be at accepting and expressing foreign genes. As the effects of a more profound grasp of chromosome structure and function become manifest, the impact on medicine and industry will prove to be far greater even than extrapolations from the current successes in the mass production by genetic engineering of rare hormones, vaccines, interferons, and enzymes.

Since the role of basic research is not always apparent to the general public, I would like to make another historical comment. Genetic engineering is solely an outgrowth of basic research. It was never planned, nor was it even clearly anticipated. Many of the procedures were discovered as unanticipated by-products of experiments designed to satisfy someone's curiosity about nature. For example, the analyses and rearrangements of DNA that form the drama of genetic engineering depend largely on a select cast of enzymes. Yet these actors were neither discovered nor created to fill these roles. Some of these enzymes, uncovered in my laboratory, came from a curiosity about the mechanisms of DNA replication. In these explorations, sponsored by the National Institutes of Health and the National Science Foundation for more than 25 years at a total cost of several million dollars, I neither anticipated nor promised their industrial application. Nor did any of my colleagues with comparable, federally funded projects. Thus, the multibillion dollar industry projected by Wall Street is entirely a product of the knowledge and opportunities gained from the pursuit of "irrelevant," basic research in universities, research made possible by the investment of many hundreds of millions of dollars by federal agencies over more than two decades.

As we retrace the flow of knowledge, we see that the first two decades of twentieth-century medical science were dominated by the microbe hunters. Their place in the spotlight was superseded for two decades by the vitamin hunters. They in turn were succeeded by the enzyme hunters in the 1940s and 1950s. For the past two decades the gene hunters have been in fashion. To whom the remaining years of our century will belong is uncertain. The neurobiologists, call them the head hunters, may very well claim it. If so, we will again see how chemistry is the fundamental language. Although brain chemistry may be novel and very complex, it is expressed in the familiar elements of carbon, nitrogen, oxygen, and hydrogen, of phosphorus and sulfur that constitute the rest of the body. Brain cells have the same DNA that all cells do; the basic enzyme patterns are those found elsewhere in the body. It is now known that hormones once thought to be unique to the brain are produced in the gut, ovary, and other tissues. The form and function of the brain and nervous system must ultimately be explained in terms of chemistry. The repeated failures of science to analyze social, economic, and political systems should not discourage us from pursuing the idea that individual human behavior, at least, can be explained by physical laws.

I sense in the future a better awareness that life can be

described in rational terms and a furtherance of chemical language to express it. For chemistry is a truly international language. It links the physical and biological sciences, the atmospheric and earth sciences, the medical and agricultural sciences. The chemical language is a rich and fascinating language that creates images of great aesthetic beauty. I see the language of chemistry taught and used for the clearest statements about our individual selves, our environment, and our society. Such visions excite me. I hope you share them. They give us courage to face the future.

Arthur Kornberg

Stanford University

CONTENTS

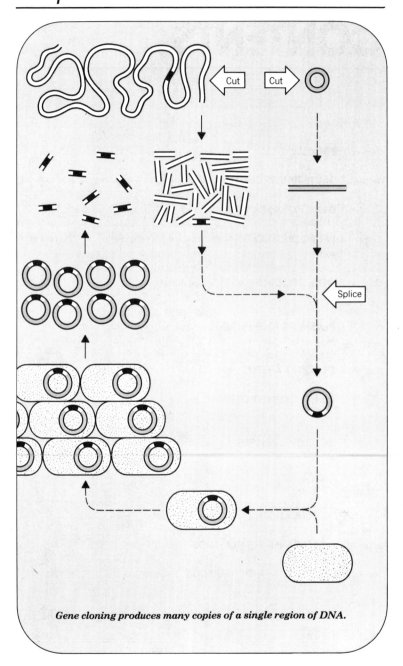

Gene cloning produces many copies of a single region of DNA.

chapter 1

PREVIEW

DNA, genes, cloning strategies, and public safety

overview

Information governing the characteristics of all organisms is stored in long, thin molecules of deoxyribonucleic acid (DNA). DNA molecules are divided into regions called genes, and genes control specific aspects of cellular chemistry. Methods are now available that allow biologists to cut any DNA molecule into specific fragments and to transfer individual fragments of DNA into a bacterium, a unicellular microorganism. A single bacterial cell receiving a specific fragment of DNA will divide repeatedly and form a cluster comprised of millions of identical cells. Each cell in a particular cluster, or clone, will contain many copies of the same DNA fragment. Billions of identical cells can be grown from a clone, and biologists can extract from these cells large amounts of a small region of DNA. That DNA is used for further study or possibly someday for correction of genetic diseases.

Since microorganisms receiving DNA fragments are being changed in unknown ways, gene cloning experiments have been viewed as potentially dangerous. Regulations were therefore established to minimize the potential health hazards. Many types of recombinant DNA have now been constructed and studied; no harmful effects have been observed, and most molecular biologists now consider cloning of genes using nontransmissible plasmids to be safe (see Appendix II).

INTRODUCTION

In a general sense biologists have solved the riddle of heredity, the question of why offspring resemble their parents. We can now explain how heredity works by describing the chemical behavior of submicroscopic structures called molecules. At the center of this new understanding is a giant molecule called deoxyribonucleic acid (DNA). This book is about DNA, the chemical that specifies features such as eye color and blood type. DNA influences all our physical characteristics as well as those of every living organism on earth. This book is also about genetic engineering, recombinant DNA, and gene cloning. In particular, it is about how gene cloning works and what has been learned from it.

To begin, **cells**, **DNA**, and **gene cloning** must be described. (All boldface words are included in the glossary.) We can think of a cell as a self-reproducing bag of chemicals and microscopic structures. Our bodies are collections of trillions of cells working together. Each cell has its own identity and function. For example, liver cells cluster together to form livers, and skin cells attach to each other to cover our bodies. It is important to note that every cell contains all the components required for an independent existence; under the right conditions it is possible to isolate a human cell and get it to grow and divide in a test tube.

DNA is one of the components required for independent existence. It is a long, thin fiber that stores the information necessary to control the chemistry of life. DNA fibers are found in every cell of our bodies, and they dictate how a particular cell behaves. Thus, DNA controls our body chemistry by controlling the chemistry of each of our cells.

Isolated DNA looks like a tangled mass of string (Figure 1-1);

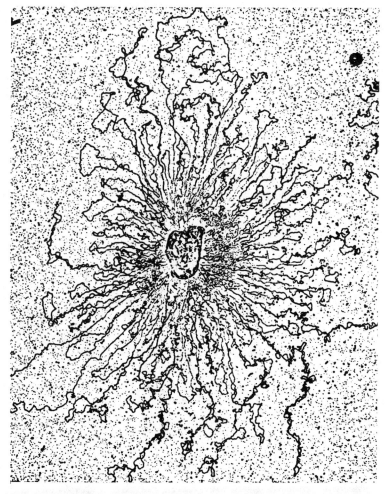

Figure 1-1. Electron Micrograph of a Single DNA Molecule Released from a Bacterium. *The long threadlike material is DNA and the dark mass in the center is the remnant of the bacterial cell wall. The entire DNA molecule is about one **millimeter** long. Photograph reprinted from* Chromosoma *by permission of Ruth Kavenoff, Oliver Ryder, and Springer-Verlag.*

each of our cells, which are generally less than a **millimeter** long, contains about two meters of DNA. Chemical analyses allow many detailed statements to be made about the structure of DNA, and in Chapters 3 and 4 some of these details are described. One of the goals of this book is to explain how this cellular information is stored, used, and reproduced. For now, the important concept is that a long, fibrous DNA molecule is subdivided into specific stretches or regions called **genes**. Each gene is responsible for causing a certain component to be made in the cell. The components interact to produce what we call life. Sometimes we can easily see the effects of particular genes; for example, a small group of genes is responsible for determining eye color. It is the specific information in the DNA, in the genes, that makes humans different from honey bees or fir trees. Information in your DNA makes you different from anyone else on earth—unless you have an identical twin.

Each DNA fiber is a **molecule**, a group of **atoms** joined together to form a distinct unit. Four points are important for understanding atoms and molecules. First, all forms of matter are composed of atoms. For example, water, wood, and steel are formed by the joining of **submicroscopic** building blocks called atoms. Second, the number of different kinds of atoms, or elements, is small (about 100). Atoms differ in size and in how they join with each other. Some are very reactive while others are inert. Hydrogen, for example, is very reactive; the airship *Hindenburg* would have been much safer if a nonreactive atom like helium had been used for lift. Third, only certain combinations of atoms join to form molecules. Thus molecules have discrete sizes and properties. Fourth, molecules can join with each other or with single atoms to form new combinations of atoms (i.e., new molecules). Such interactions are called chemical reactions. Chemical reactions are constantly occurring inside our cells. In chemical terms DNA is a giant molecule, often containing billions of atoms linked together. Information is stored in DNA by the specific arrangement of atoms, and this arrangement controls the chemical reactions in each of our cells.

Gene cloning developed in the mid-1970s when it became

possible to cut DNA and to transfer particular pieces of DNA, containing *specific* bits of information, from one type of organism into a second type of organism. As a result, the characteristics of the second, recipient organism can be changed in a *specific* way. When the recipient organism is a microbe, such as a single-celled **bacterium**, the specific fragment of transferred DNA is multiplied many times as the recipient microbe multiplies. Millions of *identical* cells, that is a **clone** of cells, eventually arise. Pure clones of bacterial cells can be easily obtained (Chapter 2); consequently, it is possible to obtain millions of copies of a specific region of DNA by placing the piece of DNA inside a bacterial cell and allowing the cell (and the piece of DNA) to multiply millions of times. This process is called gene cloning, and it is one form of genetic engineering. The potential uses of genetic engineering are many, ranging from the construction of new breeds of plants to the replacement of defective genes in humans suffering from genetic diseases.

OUTLINE OF A GENE CLONING PROCEDURE

Understanding gene cloning requires two types of knowledge: a grasp of the concepts of molecular biology and a familiarity with laboratory manipulations. Both areas are discussed in subsequent chapters. To help keep the details in perspective, the major steps in gene cloning are briefly outlined below and sketched in Figure 1-2.

The first step is to break open living cells. A number of methods are available to accomplish this. One popular way is to shear the cells in a blender and then treat them with a detergent. The next step (step 2 in Figure 1-2) is to remove genetic information from cells. This turns out to be a fairly easy, straightforward process because the information is stored in a chemical form as part of DNA. Since DNA molecules are thousands of times longer than most other large molecules found in cells, it has been possible to develop techniques for

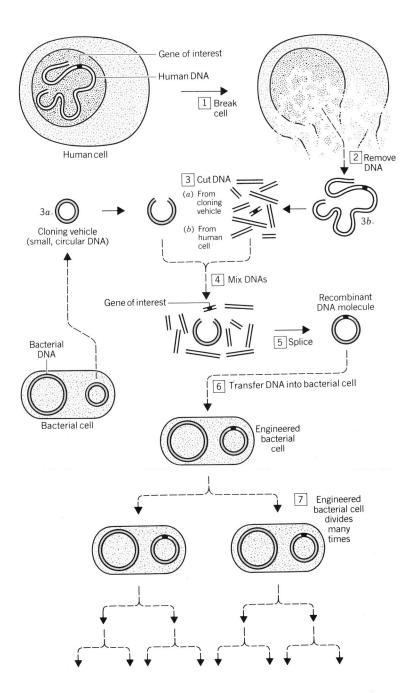

Gene of interest

Human DNA

Human cell

1 Break cell

2 Remove DNA

3 Cut DNA
(a) From cloning vehicle
(b) From human cell

3a.

Cloning vehicle (small, circular DNA)

3b.

4 Mix DNAs

Gene of interest

Recombinant DNA molecule

5 Splice

Bacterial DNA

Bacterial cell

6 Transfer DNA into bacterial cell

Engineered bacterial cell

7 Engineered bacterial cell divides many times

purifying DNA. In one method the DNA molecules are spooled onto a glass rod. The glass rod bearing the DNA molecules is then lifted out of the mixture of broken cells in the same way you would handle a fork full of spaghetti.

The third step is to cut specific genes of interest away from the rest of the DNA. In some ways the process is like editing motion picture film, for like film, DNA is divided into "frames" that make sense when seen in the correct order. In DNA the "frames" correspond to the letters in the genetic code, which are described in Chapter 3. When a number of frames or genetic letters are organized in a specific combination, they create a scene in the case of film (Figure 1-3) and a gene in the case of DNA. The molecular scissors used to cut DNA into gene-sized pieces are described in Chapter 5.

The next step is to splice these specific sections of DNA into agents called **cloning vehicles** that carry the DNA sections into other living cells (steps 4 and 5, Figure 1-2). Cloning vehicles are relatively short DNA molecules that can penetrate the wall of a living cell and can multiply inside that cell. Placing a specific gene into a cloning vehicle is equivalent to splicing a scene into a short film. The splicing process produces a **chimeric DNA molecule**, part specific gene and part cloning vehicle (Figure 1-2). Such a DNA molecule is also called a **recombinant DNA molecule.**

Once a human gene has been spliced into a cloning vehicle, both the vehicle and the human gene are transferred into a cell that is normally a **host** for the vehicle (step 6, Figure 1-2).

Figure 1-2. Major Steps Involved in Gene Cloning. *(1) Human cells are broken (for clarity only one cell is shown). (2) DNA containing the gene of interest is removed from human cells. (3) The DNA from a cloning vehicle and human DNA are cut in specific places. The cloning vehicle DNA is obtained from bacterial cells. (4) The two types of DNA are mixed. (5) The DNA fragments are spliced together, yielding a recombinant DNA molecule. (6) the recombinant DNA molecule is transferred into a bacterial cell, which has its own DNA (see Figure 1-1). (7) The engineered cell created by step 6 is allowed to reproduce millions of times to form a clone of identical cells.*

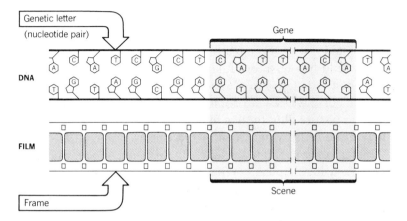

Figure 1-3. Comparison of DNA and Motion Picture Film. *The frames of movie film correspond to genetic letters (**nucleotide pairs**) in DNA. When properly organized, the frames form a scene in film and the genetic letters form a gene in DNA. DNA contains many genes, and each one stores information affecting a chemical process that occurs in living cells. Genes are generally hundreds to thousands of nucleotide pairs long. For illustrative purposes only a part of a gene is shown in the figure; slashes have been drawn through the DNA and film to indicate that many nucleotide pairs and frames have been omitted.*

Usually the host cells are single-celled organisms such as bacteria or **yeast**.

The final step in gene cloning is to allow the host cell to multiply, forming a clone having millions of identical cells. In the example shown (step 7, Figure 1-2) each member of the clone contains, in addition to its normal DNA content, the same specific piece of human DNA joined to cloning vehicle DNA. By this process a piece of genetic information, a human gene, can be transferred into a cell where it would never occur naturally. In a very limited sense, a new organism can be created.

In general, simply cloning a piece of DNA won't do much good. Just as film must be projected to be useful, the information in DNA must be converted into a useful product. To make a product, the information in DNA is usually transferred from the gene to the site where a new **protein** molecule is manufactured.

Formation of products from information stored in DNA, a process called **gene expression**, is described in Chapter 3. At that point protein structure is also discussed. For now it is important to know that proteins are chainlike molecules often containing hundreds of links. Some types of protein function as building blocks to provide cellular structure, while other types control the chemical reactions in the cell.

Insulin is a controlling type of protein, and it serves as a good example to illustrate one of the uses of recombinant DNA technology. The insulin gene is a region in our DNA that contains information for producing insulin. Some diabetics fail to produce sufficient quantities of insulin, and they are unable to properly control their **sugar metabolism**; consequently, these diabetic patients must take daily injections of insulin. Before the development of gene cloning, insulin could be obtained only by an expensive process of extracting the protein from hog pancreas; but now, through gene cloning techniques, human insulin genes have been placed in bacteria. Here the genes are expressed; that is, insulin is made inside bacteria. Thus, large quantities of insulin are now produced by bacteria, and it is easier to obtain insulin from bacteria than from pancreas tissue. Moreover, the engineered bacteria produce human insulin, an important feature for diabetics who have become allergic to hog insulin.

In summary, gene cloning is a strategy for transferring *small* bits of genetic information (DNA) from one organism to another. Certain pieces of DNA will permanently alter the chemistry of the recipient organism in useful, predictable ways.

THE SAFETY CONTROVERSY

Shortly after the first gene cloning experiments were completed in the early 1970s, scientists realized that this type of genetic manipulation might pose health hazards. No danger had been demonstrated, but one could imagine a number of scenarios frightening enough to make good science fiction copy. Suppose, for example, that the gene containing the information for botulism **toxin** were placed inside a harmless bacterium and that large numbers of this new, toxin-producing bacterium were accidentally released into the environment. A few of these organisms might find their way into the digestive tracts of humans. There they would multiply, for the human digestive tract is one of the normal habitats for the type of bacterium most commonly used in gene cloning experiments (but not for the organism that normally produces botulism toxin). If the botulism toxin were produced by the "engineered" bacteria, anyone infected with these bacteria would soon die. Little was known about the ecological relationships between common laboratory bacteria and man, and there was no way to determine whether such a scenario could be realized. Thus it seemed prudent to use recombinant DNA technology with caution.

The Stanford University scientists who developed gene cloning recognized the need for precautions. They tried to control the use of the cloning vehicles, the biological tools used to transfer genes, but it soon became apparent that they would be unable to do so by themselves. Gene cloning involves straightforward laboratory procedures, and the Stanford investigators knew that soon hundreds of scientists all over the world would be conducting experiments. To be effective, the control system would require much more muscle than a handful of scientists could flex. In June 1973 scientific reports about gene cloning were delivered to several hundred molecular biologists attending the annual Gordon Research Conference on Nucleic Acids. Concern grew among scientists, and as a result, the National Academy of Sciences established a committee to study the problem. The committee quickly recommended that an international conference be convened to formulate safeguards. It was hoped that the safeguards would be backed by the peer pressure necessary to obtain compliance. At the same time the committee publicly recommended postponement of some types of experiment.

Peer pressure is a powerful force in the scientific community; as a result, the recommendations of the committee were taken very seriously. Likewise, the recommendations of the subsequent international conference held at Asilomar, California, were incorporated into an effective system of self-regulation. In the United States the regulatory system was set up by the National Institutes of Health (NIH), the federal agency responsible for funding most recombinant DNA research. A set of guidelines was established that focused on containing recombinant organisms inside laboratories. Specific containment procedures were required for the various experiments, depending on the risks thought to be involved. In some cases the investigators could use standard laboratory techniques, whereas in others the experiments had to be conducted in specially constructed rooms isolated from the environment. Still other experiments were simply disallowed. The responsibility for compliance rested on each institution involved in recombinant DNA research. At stake was not only the health of the local community but also the institution's access to federal research grants. The institutions had to organize local biohazards committees to oversee the procedures used by each of their laboratories. By now many thousands of experiments have been performed using the NIH guidelines, and the earlier predictions of catastrophic consequences appear to have been unfounded.

Since there is no way to ensure that gene cloning is absolutely without danger, it is not surprising that use of this investigative tool became controversial. On the one hand, many argued that the risks were far outweighed by the potential benefits; that is, the risks, which were still only speculative, were well worth taking. Opponents of gene cloning, on the other hand, argued that an accident could be catastrophic. Moreover, they pointed out that the central issue of the controversy was the consent of those at risk. For example, few people would have denied Christopher Columbus and his crew, or Madame Curie, the opportunity to undertake hazardous explorations, for in these cases only the lives of the explorers were at risk. But a recombinant DNA accident, like a nuclear reactor accident, could involve many people who did not consent to be at risk. Thus some controversy will remain until enough data have been accumulated to eliminate all sense of danger.

Most scientists already view the risks as very small, and they often stress the following points. First, recombinant DNA, as an isolated material in a test tube, is not dangerous; only under carefully controlled conditions can it be transferred into living cells. Second, recombination (i.e., the forming of new combinations of DNA segments) occurs in nature and is not in itself dangerous. Third, bacteria do not often escape from laboratories and establish infections. Even highly evolved **pathogens** (disease-causing organisms) have been successfully contained in laboratories, and the isolated cases of laboratory infections have not led to epidemics. The strains of bacteria used for gene cloning do not survive well outside the laboratory; most of them are genetic cripples, often requiring special nutrients for growth. Fourth, the chance is small that one of these bacteria would be converted accidentally into a pathogen. Pathogenic organisms have sophisticated mechanisms for infecting their hosts. These mechanisms involve a number of different genes that have been carefully honed by millions of years of evolution. For all these reasons, prudent caution has been deemed adequate to prevent recombinant bacterial strains from escaping from laboratories.

An entirely different issue is whether we have the right and wisdom to manipulate our own genetic information and to influence the evolution of our own species as we have done with livestock for centuries. This is not a scientific question, and, as stated, it is generally not addressed by molecular biologists. The current thrust is to correct genetic defects in particular individuals, not in the species in general. For example, individuals having genetic defects in blood cells may be treated by changing the DNA in their blood cells, not in their **germ cells** (**sperm** and **eggs**). Thus the individual, not his or her children, will be affected by genetic engineering. It is conceivable, however, that knowledge obtained from recombinant DNA technologies will make it possible to modify germ cells. Then it may be up to you as a citizen to decide how or indeed whether this type of knowledge will be applied.

PERSPECTIVE

To understand gene cloning, it is necessary to become familiar with each of the steps in the process. It's not quite as simple as Figure 1-2 suggests. As an example of the complexity, consider the following point: the molecular "scissors" used to cut DNA during the editing process (Figure 1-2, step 3) makes many, many cuts, producing thousands of different pieces of DNA. Locating the single, desired fragment of DNA and separating it from all the other pieces is a formidable task. Unfortunately, DNA cannot be run through an editing machine like a piece of film; genes in DNA are too small to be seen by the human eye. Even with the highest powered microscopes, DNA containing thousands of genes looks like a featureless piece of string (see Figure 1-1). Instead, gene cloners blindly separate DNA fragments; then they examine the different fragment types biochemically until they locate the desired one. The separation process involves putting individual DNA fragments *randomly* into many, many bacterial cells. The next chapter describes features of bacterial growth that facilitate separation of DNA fragments. Chapter 3 discusses how information in DNA is converted into useful products. Chapter 4 outlines how new DNA molecules are formed inside living cells and introduces some of the protein tools used in gene cloning. At that stage enough background material on DNA structure will have been presented for us to return to the beginning of the cloning procedure and consider in detail how DNA fragments are spliced into cloning vehicles. At the same time, methods for retrieving cloned genes from bacterial cells will be described. By Chapter 7 it will be possible to tie all the individual steps together by describing how hemoglobin genes were first cloned. Chapter 8 focuses on the use of cloned genes for determining the organization of our genetic information, and the final chapter introduces a few of the more significant findings derived from gene cloning. By that point, you will have a general understanding of one of the major strategies biologists are using to discover how living cells work.

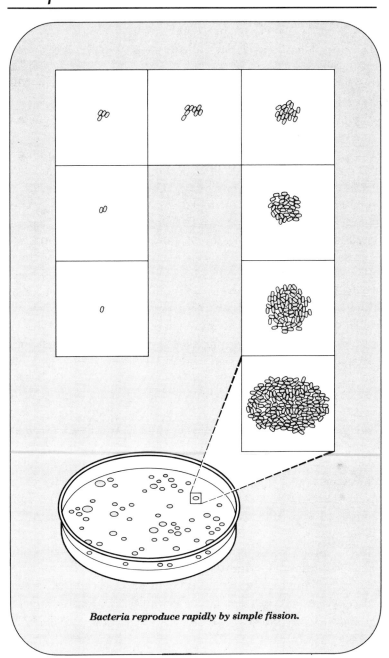

Bacteria reproduce rapidly by simple fission.

BACTERIAL GROWTH

microorganisms as tools for gene cloning

overview

Bacteria are microscopic organisms that reproduce very rapidly and are easily grown in the laboratory. Thus they are useful for studying how traits are passed from one generation to the next. A single cell placed on a solid surface containing nutrients can multiply to form a colony comprised of millions of identical cells. If the colony is transferred to liquid growth medium, the cells continue to multiply; within one to two days it is possible to obtain a culture containing trillions of identical cells.

Bacteria are used in two ways for genetic engineering. First, they are used to separate different DNA fragments. Conditions can be obtained in which a small piece of DNA will pass into the interior of a bacterial cell. The cell containing the piece of DNA can be separated from all other cells and grown until it forms a colony comprised of a million identical cells. Second, pure cultures, in which each bacterium contains the same type of recombinant

DNA, can be grown in huge quantities to produce desired protein products.

INTRODUCTION

Bacteria are one-celled organisms (Figure 2-1) that live almost everywhere: in the soil, on our skin, and in our intestines. When most people think about bacteria, diseases come to mind, diseases like plague, botulism, anthrax, cholera, and typhoid fever. Fortunately, most bacteria are not pathogenic. Some of the nonpathogenic species have become the favorite experimental subjects of molecular biologists, and the study of these microbes has led to gene cloning technologies. The two sections that follow outline the growth properties of bacteria that make these microorganisms so useful. The final section of the chapter briefly describes some of the life processes that occur inside bacteria to help clarify the role of DNA.

CHARACTERISTICS OF BACTERIAL GROWTH

About 40 years ago, molecular biologists became interested in bacteria because these organisms are very tiny (a high power microscope is needed to see individual bacteria) and because bacteria seem to lead such uncomplicated lives. They simply grow and then divide in half to produce two new cells. Each of the new, daughter cells then expands until it, too, divides to form two more cells. Thus bacteria reproduce by simple **fission**. This simplicity produced the hope that everything about bacterial life could be understood in molecular terms, that the essence of life itself could be understood.

Two properties of bacteria are particularly important. First, each bacterium is only a single cell, lacking limbs, organs, and complicated developmental stages. Consequently, it is relatively

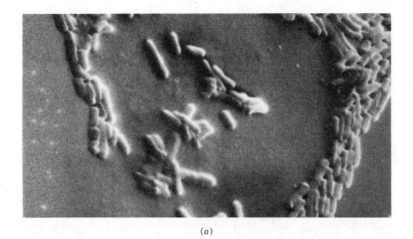

(a)

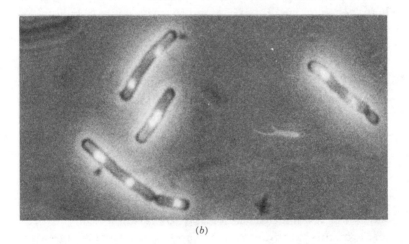

(b)

Figure 2-1. Photomicrographs of Bacteria. *(a) A cluster of E. coli cells as they appear using scanning electron microscopy. Magnification is 3800 times. Photomicrograph courtesy of Sandra McCormack, Rochester Institute of Technology. (b) Several E. coli cells as they appear using light microscopy. The bright structures in the center of the cells are DNA-containing bodies called* **nucleoids***, which have been stained with a dye called ethidium. An electron micrograph of a nucleoid that has been removed from the cell is shown in Figure 1-1. The cells shown here are mutants that are unable to reproduce properly. As a result, the cells become elongated and appear to have more than one nucleoid per cell. Magnification is 3800 times. Photomicrograph courtesy of Todd Steck, University of Rochester.*

easy to obtain large numbers of a *single type* of cell to study its chemistry. Second, bacteria grow and multiply rapidly. Experiments that would take years with other organisms can be done in a single day with bacteria. For example, in the laboratory many kinds of bacteria divide every 40 minutes. Thus, during an 8-hour day, a batch of reproducing bacteria will go through 12 generations, greatly facilitating the study of how traits are passed from generation to generation. Comparable experiments with humans would take more than 200 years.

The two properties mentioned above, small size and rapid growth, make it easy to cultivate large numbers of bacteria. The recipe for growing bacteria is simple: place a few bacteria in lukewarm **broth**, and let the bacteria do the rest. They grow and divide, producing what is called a **bacterial culture**. Within 24 hours even a small flask of broth may contain billions of bacteria. Thus it is easy to get astronomical numbers of bacteria.

Although the growth properties of many bacteria types make them suitable tools for gene cloning, a species called *Escherichia coli* (**E. coli** for short) is used most extensively. *E. coli* is not particularly distinctive; it is a small rod-shaped organism (Figure 2-1) that is normally a harmless inhabitant of the human digestive tract. Like most other bacteria, its shape is maintained by a rigid coating called a **cell wall**. Inside this wall around the cell occur the chemical reactions of life. *E. coli* is special because 35 years ago a group of molecular biologists focused their research on it. As details about the chemistry of *E. coli*'s life were learned, it became easier to conduct more sophisticated experiments on this bacterium than to start over with a new organism. As a result of this intense effort, *E. coli* has become the best understood organism on earth. Likewise, the best known cloning vehicles are infectious agents that attack *E. coli*, for they too have been intensively studied for many years.

Since our cloning tool, *E. coli*, is so small, special methods are required to use it. One cannot simply look through a microscope to see whether a particular cell has taken up a specific piece of DNA; the cells have few distinctive characteristics when viewed through an ordinary microscope. Only under special conditions can the DNA be seen inside the cell, and then it looks like a featureless blob (Figure 2-1*b*). Nor can one dissect a bacterium, as many of us dissected frogs in general biology class. Instead, indirect observations must substitute for what cannot be seen. For example, a liquid bacterial culture will appear cloudy if it contains more than 10 million cells per cubic centimeter (a quarter-teaspoon). As the bacteria multiply, the cloudiness increases. Thus, by measuring how fast the cloudiness increases, it is possible to determine the rate at which the bacteria are reproducing.

BACTERIAL COLONIES

A single bacterial cell growing on a solid surface will multiply to form a cluster of cells called a **colony**. Since this is one of the more important concepts in gene cloning, a procedure is presented below for obtaining bacterial colonies, a procedure that could be carried out in almost any kitchen. Biologists use slightly more refined equipment to obtain colonies, but the principles are the same. First **dissolve** in boiling water some **agar**, which is a gelatinlike substance. Pour the resulting solution into a **sterile**, covered dish and set aside, allowing the agar to cool and solidify. Next place a teaspoon of soil in a tablespoon of water. Stir briefly. Since many types of bacteria live in soil, it is a good place to obtain these organisms. Bacteria such as *E. coli* can be obtained from sewage. Now find a thin piece of wire and bend the end to form a loop about a millimeter in diameter. Dip the loop directly in the mixture of soil and water and lift it out. The loop will trap a tiny drop of water containing bacteria. Place the drop of water from the loop onto the surface of the solid agar and smear the drop over the surface of the agar with the wire loop. Replace the lid on the dish and allow the bacteria to grow and multiply at room temperature. Several days later hundreds of small bumps can be seen on the surface of the agar. These bumps, which often look like glistening blisters about a millimeter across (see Figure 2-2 and frontpiece for Chapter 2), are bacterial colonies.

When the drop of water was smeared over the agar, a small number of bacteria from the soil were scattered to widely separated spots on the agar. Each cell divided many times, and since the new cells could not move away from each other, they piled up. Within a day or so, the bacterial colony became visible. With this simple procedure it is easy to separate and culture the individual bacterial cells in the original soil sample. The millions of bacteria in a colony all arose from a single bacterial cell. Thus all cells in a colony are identical; they are members of a **clone** (step 7, Figure 1-2).

The ability to obtain individual colonies is important because gene cloners cannot visually distinguish one gene from

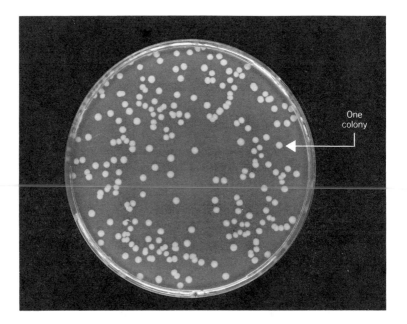

Figure 2-2. Bacterial Colonies Growing on Agar. *A dilute suspension of E. coli cells is spread on solid agar in a petri dish and is incubated at 37° C. Usually after 24 hours colonies, each about 2 millimeters in diameter, are visible on the agar. The arrow points to one such colony.*

another. We cannot simply use forceps to pick particular genes out of a pile of DNA fragments. Instead, we use bacteria as outlined in Figure 1-2. Bacteria can incorporate small pieces of DNA if the DNA is linked to a cloning vehicle, a small, **infectious** DNA molecule. Thus, gene cloners separate DNA fragments by first transferring them into bacteria and then spreading the bacteria on agar. The bacteria grow into visible colonies, and the cloned DNA fragments multiply millions of times. The colonies are then tested for the presence of particular DNA fragments as described in later chapters.

Finding the colony that contains a particular gene is not easy; the gene cloner faces a problem of numbers. First, the gene being sought may be rare; it is not uncommon for the DNA fragment containing a particular gene to represent less than one out of 100,000 DNA fragments. Second, transferring DNA into bacterial cells is an inefficient process. Only one in 10,000 cloning vehicles will take up residency in a cell, and only a fraction of these will be attached to one of the fragments being sought. Thus the chance that any particular bacterial cell will contain a specific fragment may be less than one in a billion. The cloner's main task is to find that rare cell.

Whenever biologists work with an organism, they try to obtain a pure culture, one that contains only the type of organism being studied. Interpreting experimental results or producing a pure product is very difficult if other types of organism contaminate the culture. Our ability to grow single bacterial colonies makes it easy to obtain pure cultures. The principle of pure cultures can be illustrated by describing one way to achieve commercial production of insulin with bacteria. First, human DNA is obtained as described in Chapter 7. The DNA is cut into millions of discrete fragments and spliced into cloning vehicles (step 5, Figure 1-2), producing many different types of recombinant DNA molecules. The collection of recombinant DNA molecules is next mixed with a huge number of *E. coli* cells. Some of the recombinant DNA molecules get inside bacterial cells. The cells are then spread out on the surfaces of agar plates to separate one cell from another. Colonies grow, and each is biochemically tested by strategies described in later chapters and in Appendix III until a colony containing insulin genes is found. That particular colony is touched with a piece of sterile wire in such a way that some of the bacterial cells in the colony stick to the wire. The cells on the wire are then transferred into a flask of sterile broth by simply dipping the end of the wire into the broth. Some of the cells fall off the wire and begin to grow and divide in the broth. By the next day the flask will be full of bacteria, all of which have the insulin gene. The flask contains what microbiologists call a pure culture of bacteria, one having only a single type of organism; all the cells in the flask are members of a clone. This culture can be maintained indefinitely if care is taken to keep other bacteria out of the culture. To grow large amounts of the bacteria, one simply transfers a drop from the pure culture into a vat of sterile broth. Within a few days the vat will contain trillions of bacteria all having the insulin gene.

BIOCHEMICAL ASPECTS OF BACTERIAL GROWTH

Bacterial cells contain elaborate mechanisms that allow them to reproduce. A knowledge of how this machinery works is central to understanding genetic engineering, for most of the gene cloning tools, including the genes themselves, are components of this machinery. We can begin to dissect the machinery by first focusing on the relationships between cellular **chemical reactions, enzymes**, and DNA.

Although people have many definitions of life, one experimentally useful view is that life is a collection of chemical reactions organized to reproduce itself. In these reactions molecules are broken down into simpler molecules, built up into more complicated molecules, or simply rearranged to create slightly different molecules. Over the years chemists have been able to describe a large number of the rules that govern these reactions. In some ways a series of chemical reactions is similar to tailoring a shirt: the cloth is cut into specific shapes, the front and back are joined together, the sleeves and collar are stitched, and then they are attached to the body. Finally buttons are added. Each step is analogous to one of the chemical reactions in the series; through a series of steps the original piece of cloth is converted into a new form.

Although one person working alone can make a shirt, the analogy to enzyme chemistry is clearer if we think of each task as being the assignment of a particular workman—a cutter, a stitcher, and so on. If shirts are to be produced efficiently, the tasks must be performed in a particular order. So it is with the work of enzymes, the specialized molecules that control the chemistry of cells by determining which chemical reactions occur and when they occur. Enzymes follow rules that dictate an order to the chemical reactions. In E. coli there are about a thousand DIFFERENT kinds of enzyme that control production of the cellular components. The key idea is that the chemicals in E. coli, as well as those in every cell in our bodies, react in an orderly fashion; it is the enzymes that provide that order. Some enzymes are very valuable engineering tools, particularly those that cut DNA and those that splice DNA.

Enzymes are members of the class of molecules called **proteins**, long, linear molecules that are much like beaded necklaces twisted in a specific way (Figure 2-3). Enzymes sometimes contain hundreds of beads. Each bead is called an **amino acid**. There are 20 different *kinds* of amino acid, so there are many possible ways in which a chain can be put together. Consequently, it is easy to show that there could exist thousands of different kinds of enzyme. It is important to realize that the properties of a particular enzyme, that is, how it controls a particular chemical reaction, are determined by the *order* of the amino acids in the chain. Consequently, knowing how the cell specifies the precise order of amino acids in an enzyme is central to understanding living processes. Such knowledge is also the key to understanding heredity, for the characteristics of enzymes and other proteins make a cell what it is. The information determining the order of amino acids in every enzyme (indeed, every protein) is stored in DNA (Figure 2-3).

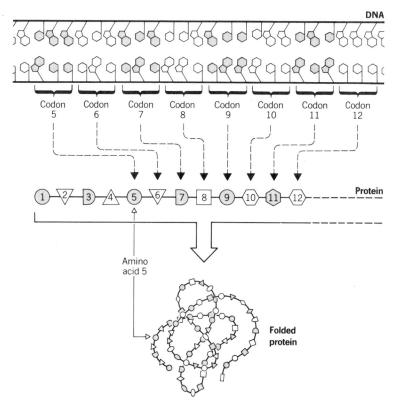

Figure 2-3. Relationship of Information in DNA to Protein Structure. *Information in DNA is arranged in a series of three-letter words called* **codons.** *Each codon specifies a particular amino acid. For example, codon 5 has information for the fifth amino acid in the protein. The overall shape and activity of a protein depend on the precise order of amino acids. The information for this order is stored in the DNA. The two DNA strands are wound around each other (see Figure 3-1); for clarity, winding of the DNA strands is not shown here.*

PERSPECTIVE

Throughout our history bacteria have plagued us by causing diseases. We have learned to live with these tiny organisms by washing our hands, treating our drinking water, immunizing our bodies, and keeping bacteria-infected fleas from biting us.

Occasionally our efforts fail, and we have to kill bacteria with antibiotics. We have also learned to use bacteria. For example, bacteria help us make products like yogurt. Recently, our relationship with bacteria has entered a new phase: we are now using them to help understand and manipulate the chemistry of life. The next four chapters summarize some of the important biological concepts we have learned from bacteria.

chapter 3

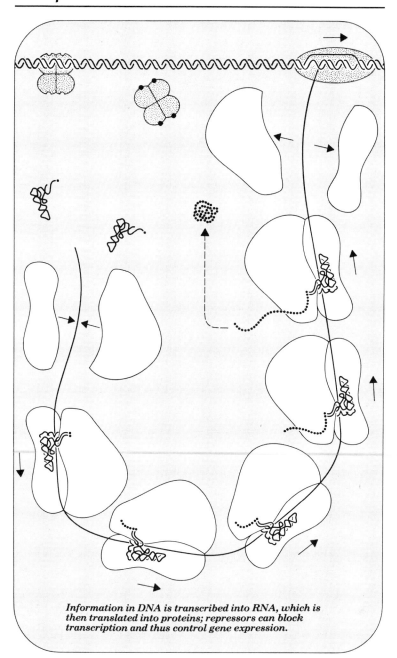

Information in DNA is transcribed into RNA, which is
then translated into proteins; repressors can block
transcription and thus control gene expression.

GENE EXPRESSION

storage and use of information

overview

A gene is a specific region of DNA, a specific stretch of nucleotides, containing information for a particular protein. Nucleotides are the subunits that compose each long, two-stranded molecule of DNA. Cells make thousands of different proteins, and each cell contains thousands of different genes. To make a certain protein, the cell selectively copies the information from the short segment of DNA that is the appropriate gene for the desired product. This copying entails the production of an RNA molecule whose nucleotide sequence accurately reflects the sequence of nucleotides of the gene itself. RNA has a structure similar to that of DNA, and both RNA and DNA use a 4 letter alphabet to store information. This type of RNA is called messenger RNA. The information in messenger RNA is then used to direct the formation of a specific protein by a process called translation. In this process messenger RNA first binds to subcellular workbenches called ribosomes.

Next, amino acids, the subunits that make up proteins, align in the order specified by the messenger RNA. During the alignment process the amino acids are linked together to form a protein chain.

The alignment of amino acids involves a second kind of RNA molecule called transfer RNA. Transfer RNA molecules solve the problem of converting the 4-letter alphabet used in DNA and RNA into the 20-letter alphabet (because there are 20 different kinds of amino acid) used in proteins. Transfer RNA molecules come in more than 20 varieties, at least one variety for each type of amino acid, and they serve as adapters. One end of a transfer RNA attaches to a specific type of amino acid while another part attaches to a specific section of the messenger RNA. Thus the adapters line up along the messenger RNA in the precise order dictated by the message from the gene. Since each adapter is also attached to one of the amino acids, ordering the adapters also orders a series of amino acids. Elaborate mechanisms have evolved to control the timing of gene expression, that is, to dictate when a given protein is made from the information stored in its gene. Successful genetic engineering requires that these regulatory mechanisms be understood.

INTRODUCTION

Since our bodies function through the combined action of trillions of tiny cells, understanding our total body chemistry requires a knowledge of how individual cells work. At the end of the preceding chapter it was emphasized that cells are composed of chemicals, many of which are constantly being converted into other chemicals. The chemical conversions occur in a series of steps controlled by chainlike molecules called proteins. The information required for the production of each protein is stored in another chainlike molecule called DNA; thus DNA indirectly controls cell chemistry by regulating the production of specific proteins. By changing the DNA in a cell, genetic engineers can permanently alter the chemistry, and thus the characteristics, of a cell. We next need to consider how information is stored in DNA and how information is converted into useful products. These questions are addressed by focusing on two aspects of DNA and genes: the chemical structure of genes and the regulation of gene expression. A detailed knowledge of DNA structure is crucial to understanding gene splicing, cloning, and detection; understanding the principles of gene expression makes it possible to appreciate the difficulties engineers face when they try to control gene function.

STRUCTURE OF GENES AND DNA

In earlier chapters the organization of information in DNA was described by drawing parallels between it and motion picture film. But to describe the chemistry of DNA, other analogies must be developed. One can think of DNA as a long, thin string composed of two strands wound around each other much like strands in a rope (Figure 3-1a). But closer inspection, using biochemical rather than microscopic methods, reveals that each strand is composed of tiny subunits; thus, the DNA strands are more accurately thought of as two interwound strings of beads. Each string contains millions of minute subunits linked together (Figure 3-1b). Chemists call the subunits **nucleotides**, and they have found that each nucleotide is composed of two parts: a flattened structure called a **base**, which points toward the other strand of DNA, and a small component of the backbone of the DNA strand (Figure 3-1c). As the backbones of the two strands wind around each other, they form a double helix (Figure 3-1c), leading to the popular expression for DNA.

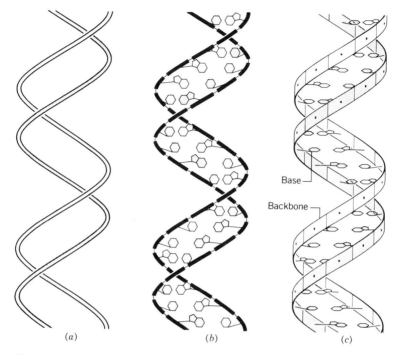

Base

Backbone

(a)   (b)   (c)

Figure 3-1.   Schematic Representations of DNA. *(a) Two interwound strands. (b) Two interwound beaded chains. (c) Double helix of two interwound strands with bases on the inside and backbone on the outside.*

The backbone of a DNA strand is composed of two alternating parts, a sugar and a phosphate. A more detailed representation of DNA (Figure 3-2) illustrates their relative orientation. The bases, which are perpendicular to the sugars, tend to stack one on top of another, much like steps in a spiral staircase. The bases come in varieties, popularly abbreviated as A, T, G, and C. The letters stand for **adenine, thymine, guanine**, and **cytosine**, the chemical names of the bases. Since each nucleotide contains only one base, the nucleotides can also be identified by the same four letters. The order of these four nucleotides is precisely arranged in DNA, and it is through this arrangement of nucleotides that cells store information. The principle is similar to the

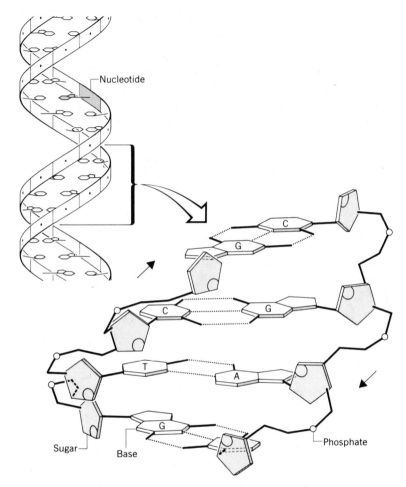

Figure 3-2. Schematic Diagram of a Short Section of a DNA Double Helix.
*Each DNA strand is composed of three types of chemical structures: bases, sugars, and phosphates. A nucleotide is a unit composed of one base, one sugar, and one phosphate. The sugars and phosphates connect to form the backbone of each strand, and a base attaches to each sugar. The four different bases are represented by the letters A, T, G, and C. The bases of one strand point inward and toward those of the other. Attractive forces called **hydrogen bonds** (represented by dotted lines) exist between the bases of opposite strands and contribute to holding the two strands together. The two strands run in opposite directions; notice how the sugars in one strand seem to point upward while those in the other seem to point downward.*

Morse code, where information is transmitted in combinations of two symbols, dots and dashes.

Extensive examination of DNA has led to the identification of two rules. First, a single DNA strand never has branches. Consequently, the information is stored in a simple line. Second, when two DNA strands come together and form a double helix, bases must fit together in a precise way. Whenever an A occurs in one strand, a T must occur opposite it in the other strand. Likewise G always aligns opposite to C. Only when the bases are properly paired will the two DNA strands fit together. This second rule is called **complementary base-pairing**. It is important to note that the two strands of DNA are complementary, NOT identical; identical nucleotides do not form base pairs. One can imagine that the bases opposite each other fit together like electrical plugs in sockets (Figure 3-3); only the correct pairs fit together. As a result of the millions of tiny plugs and sockets fitting together and the bases stacking on top of each other, the two strands of DNA tend to stick tightly together. Thus the DNA double helix is a stable structure; temperatures near that of boiling water are required to separate the strands.

The subunits of the DNA strands, the nucleotides, are the chemical basis for storage of information in DNA. Returning to the film analogy introduced in Chapter 1 (Figure 1-3), the units we have now defined as nucleotides pairs, or base pairs, correspond to the frames in a motion picture film. That is, the "genetic letters" mentioned earlier represent the chemical base name abbreviations A, T, G, and C. Information is stored in DNA in the form of a 4-letter code (A, T, G, C), which reads in a line along the DNA like frames in a film. In physical terms, a gene is a stretch of DNA ranging from a few hundred to a few thousand nucleotide pairs; it corresponds to a scene in the motion picture film.

We can begin to define a gene more precisely by considering features important for converting information from DNA into protein. First, a gene has a beginning and an end; there are short stretches of nucleotides that signal where the gene starts and

(*a*) **Structural formulas**

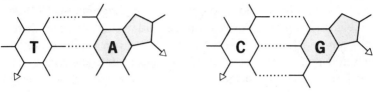

(*b*) **Prongs and sockets**

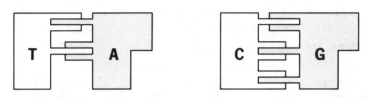

Figure 3-3. Complementary Base-Pairing. (*a*) *Structural formulas for thymine:adenine (TA) and cytosine:guanine (CG) base pairs. The bases are flat structures composed of hydrogen, carbon, nitrogen, and oxygen atoms. The solid lines represent chemical bonds between these atoms. Arrows indicate where the bases attach to sugars. The dotted lines are hydrogen bonds, weak attractive forces between hydrogen and either nitrogen or oxygen. Notice that there are two hydrogen bonds between adenine and thymine, and three between guanine and cytosine. The difference in hydrogen bonding is the structural explanation for complementary base-pairing. (b) A prongs-and-sockets analogy for base-pairing. The hydrogen atoms in each hydrogen bond are represented as prongs, and the oxygen or nitrogen atoms are depicted as sockets. The attractive forces are weak; consequently, perfect fits are required for base-pairing to occur.*

other short segments that indicate where it stops. Second, the information in a gene is arranged as words rather than as individual letters. This is because proteins, which are also linear chains, are composed of 20 kinds of subunit (amino acids) that are chemically different from the 4 subunits (nucleotides) used to store information in DNA. A quick calculation predicts how many nucleotide letters would be necessary to code for each amino acid, that is, how many letters in DNA correspond to each word in a protein. Obviously one nucleotide in DNA cannot correspond to one amino acid in protein because there are 20 different amino acids and only 4 different nucleotides. Likewise, the nucleotides cannot be read in sets of two because 4 different nucleotides taken two at a time can produce only 16 possible pairs, 4 short of the minimum number. The code can be read in threes: 4 nucleotides taken three at a time give 64 possible triplets. This is more than enough to specify the 20 amino acids as well as the necessary punctuation, such as start and stop signals. Many experiments have confirmed the prediction that the genetic code is read as triplets of nucleotides. Triplets in DNA that correspond to amino acids are called **codons**. For example, the codon T–C–A specifies that the amino acid called serine be placed in a protein chain.

There is no punctuation between the codons. Thus it is important that the reading frame of the nucleotide code be established correctly—the start signal must be in the right place. This principle can be illustrated by the following sentence read as 3-letter words:

JOE SAW YOU WIN THE BET

If you read in sets of three and start out of register at the letter *O* instead of *J*, you end up with a meaningless sentence:

OES AWY OUW INT HEB ET

Consequently, whenever an engineer splices genes into a new DNA molecule, the correct start signal must be present to establish the right reading frame, both for the amino acids being joined and for the stop signal.

GENE EXPRESSION: TRANSCRIPTION

Gene expression is the process of producing a protein from the information stored in DNA. This process involves a number of steps and components. First, the stored information from specific regions of the DNA is made available to the rest of the cell through a process called **transcription** (Figure 3-4). In this process an enzyme, **RNA polymerase**, recognizes and binds to a DNA nucleotide sequence near one end of a gene. This recognition and binding allows RNA polymerase to select the proper DNA strand (the strands are complementary, not identical) to use as a source of information. The polymerase then moves into the gene. As RNA polymerase moves, it creates a new chain by linking together RNA nucleotides floating free in the cell. The order of the nucleotides in the new chain is determined by the complementary base-pairing rule. If the first letter the RNA polymerase encounters in the DNA is a T, the enzyme will add an A to the new chain it is making. Likewise, if the next DNA letter is G, a C is added to the new chain. Eventually a stop

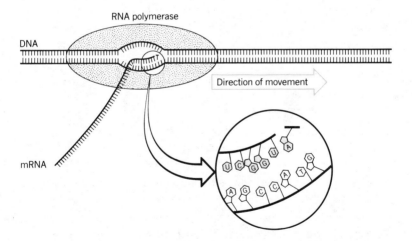

Figure 3-4. Transcription. *The enzyme complex called RNA polymerase causes the DNA strands to separate within a short region (10 to 20 base pairs). The polymerase moves along the DNA, and as it does, it forms an RNA chain using free nucleotides. The order of the nucleotides in RNA is determined by the order of nucleotides in one of the DNA strands by the complementary base-pairing rule. In the example shown, the nucleotide sequence of the RNA is complementary to that of the lower DNA strand. For simplicity, the DNA strands have not been drawn as an interwound helix (see Figure 3-1).*

signal at the end of the gene is reached, RNA polymerase comes off the DNA, and the new chain is released. This new chain is called **RNA**. RNA is similar to DNA in having only four types of subunit (nucleotides). However, RNA is a shorter chain; it generally occurs as a single strand with a slightly different backbone, and it contains the base **uracil** (U) instead of thymine (T).

GENE EXPRESSION: TRANSLATION

Once the information from a gene has been transcribed into an RNA molecule, the RNA serves as a messenger, transporting the information from the DNA to subcellular structures called **ribosomes**. Ribosomes are large (in molecular terms) ball-like structures, and it is here that the information in the messenger RNA is **translated** from the nucleotide language into the amino acid language. As a chain of amino acids is made, it *spontaneously* folds to form the protein specified by the gene in the DNA.

The translation machinery works in the following way. Messenger RNA attaches to a ribosome near a site on the messenger called the start codon, a three-base triplet that indicates where to start reading the message (Figure 3-5). In bacteria this begins before the messenger RNA synthesis has been completed; thus, messenger RNA and ribosomes can be attached to DNA (Figure 3-5); in our cells the messenger RNA leaves the DNA before attaching to ribosomes. Meanwhile, the 20 different types of amino acid, which eventually will be linked to form a protein, are floating free in the cell. Each amino acid is joined to another type of RNA molecule called **transfer RNA** (tRNA). Transfer RNA functions as an adapter to read the information on the messenger RNA. To accomplish the joining, each type of amino acid is recognized by a special type of enzyme (Figure 3-6a) called an **aminoacyl-tRNA synthetase**. There are more than 20 different aminoacyl-tRNA synthetases,

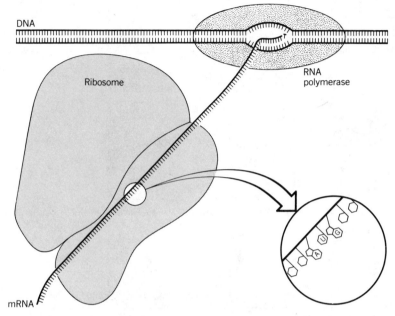

Figure 3-5. Schematic Representation of Messenger RNA Being Formed by RNA Polymerase and Attaching to a Ribosome. *The ribosome is composed of two ball-like structures that bind to messenger RNA. Ribosome–messenger binding requires that a particular transfer RNA also bind to the A-U-G (or in some instances G-U-G) codon on the messenger mRNA. This transfer RNA (not shown) is attached to the amino acid destined to become the first in the new protein chain (see Figure 3-7). In bacteria, the messenger RNA is still attached to DNA when it binds to a ribosome. In more advanced organisms, such as humans, the messenger RNA is released from the DNA before attaching to a ribosome.*

at least one for each type of amino acid; each of these enzymes recognizes and attaches to only one of the 20 types of amino acid. Each enzyme is also able to recognize and attach to a specific type of transfer RNA. There is a different type of transfer RNA for each type of amino acid. Once a particular amino acid and a particular transfer RNA have attached to a particular aminoacyl-tRNA synthetase, the synthetase links the amino acid to the transfer RNA (Figure 3-6*b*). The amino acid–

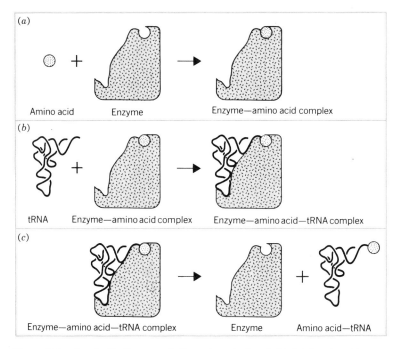

(a)

Amino acid   Enzyme   Enzyme—amino acid complex

(b)

tRNA   Enzyme—amino acid complex   Enzyme—amino acid—tRNA complex

(c)

Enzyme—amino acid—tRNA complex   Enzyme   Amino acid—tRNA

Figure 3-6. Symbolic Representation of an Amino Acid Joining to a Transfer RNA. *(a) An aminoacyl-tRNA synthetase (labeled "Enzyme") recognizes and attaches to an amino acid. Each of the 20 different amino acids is recognized by a different aminoacyl-tRNA synthetase. (b) The enzyme then recognizes and binds to a specific type of transfer RNA (there are more than 20 different types of transfer RNA molecules, at least one of each type of amino acid). In the process the transfer RNA and the amino acid are joined to form what is called an aminoacyl-tRNA. (c) The aminoacyl-tRNA is released from the enzyme, which is then free to repeat the process.*

transfer RNA pair is then released from the enzyme (Figure 3-6c). The net effect is to create 20 different amino acid–transfer RNA pairs.

Each of the transfer RNAs has a three-nucleotide region called an **anticodon** (C–A–U in Figure 3-7) located opposite the end where the amino acid is attached. Each of the 20 types of transfer RNA has a different anticodon. Thus the particular amino acid at one end of the transfer RNA always corresponds

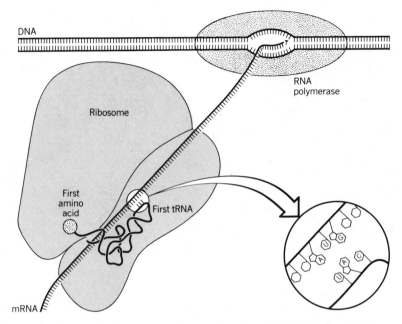

Figure 3-7. **Recognition of Codons by Anticodons.** *The start codon (A-U-G) on the messenger RNA and the anticodon (C-A-U) of the first transfer RNA bind on the ribosome. The amino acid destined to be first in the new protein chain is already attached to the first transfer RNA. RNA and DNA have distinct left and right ends, and by convention nucleotide sequences are always written left to right; however, when base-pairing occurs, one strand runs left to right and the other runs right to left (see Figure 3-2). Consequently, if the codon is written as A-U-G, the anticodon must be inverted during base-pairing and appears as U-A-C in the drawing.*

to a specific set of three nucleotides in the anticodon region of the transfer RNA. The specificity of the aminoacyl-tRNA synthetase ensures that this is the case. The anticodon on the transfer RNA can be exposed to form base pairs with the messenger RNA (Figure 3-7). One particular transfer RNA has an anticodon triplet complementary to the start codon, or triplet, on the messenger RNA. That transfer RNA and the messenger RNA lock together on the ribosome so that the two triplets, the codon on the messenger and the anticodon on the

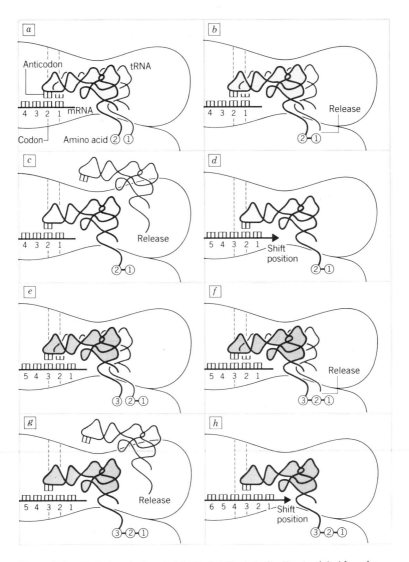

Figure 3-8. Ordering Amino Acids During Protein Synthesis. *(a) After the messenger RNA (mRNA), initiator aminoacyl-tRNA, and ribosome have formed a complex (Figure 3-7), a second tRNA, with its attached amino acid, is ordered on the ribosome when its anticodon region base-pairs with the second codon of the mRNA. Dashed lines are reference lines to show positioning of codons on the ribosome. (b) Amino acids 1 and 2 are bound together; ami-*

transfer RNA, form base pairs. This joining is governed by the complementary base-pairing rule: if the start codon on the messenger RNA is A–U–G, the only transfer RNA that will fit has an anticodon that reads C–A–U (if you flip C–A–U over, it forms U–A–C, which is complementary to A–U–G; see legend to Figure 3-7). The particular amino acid attached to this transfer RNA is destined to become the first link in the new protein chain.

Figure 3-8 illustrates how the amino acids are ordered in the new protein. The second triplet codon on the messenger is also locked into place on the ribosome next to the first codon. It too is recognized by the anticodon of a transfer RNA molecule carrying an amino acid, the amino acid destined to become the second link in the new protein. Other proteins attached to the ribosome then join the two amino acids together. The first amino acid separates from its transfer RNA, and that transfer RNA separates from the messenger, completing one cycle of the translation process. The messenger now feeds across the ribosome much like a magnetic tape over the player head of a tape recorder. One after another the triplet codons are locked into place on the ribosome. The appropriate transfer RNA binds to each triplet, placing the correct amino acid in position to be joined to the growing protein chain. When the stop signal comes along, the messenger falls off the ribosome, and the new protein is released into the cell, where it begins to control the specific chemical reaction it was designed for.

In summary, information from the DNA, **encoded** by four

←————————————————————————————————

no acid 1 is released from tRNA 1 (note break). (c) tRNA 1 is released from the ribosome. (d) mRNA and tRNA 2, now attached to two amino acids, are translocated (shifted over one position on the ribosome). This brings codon 3 into position on the ribosome. (e) Aminoacyl-tRNA 3 attaches to the ribosome and forms base pairs with eodon 3. (f) Amino acid 3 binds to amino acid 2, repeating step b. (g) tRNA 2 is released from the ribosome, repeating step c. (h) tRNA 3 and the growing protein chain are translocated, repeating step d. This process continues until a stop codon is reached. At this point the protein chain is released from the last tRNA.

different letters, is first transcribed into a message (messenger RNA), using a similar 4-letter code. The message then binds to a ribosome and is translated into a protein chain. Small RNA molecules, called transfer RNAs, serve as adapters to convert the 4-letter alphabet of DNA and RNA into the 20-letter alphabet of proteins. In so doing, the transfer RNA molecules move amino acids floating free in the cell into position on ribosomes where the amino acids are linked together to form a protein chain. All organisms on our planet use this process for making proteins, leading biologists to conclude that life is a continuum.

CONTROL OF GENE EXPRESSION: REPRESSION

Whether genes are to be placed in bacteria to produce large quantities of a specific protein, into animals or humans to correct defective genes, or into plants to improve food sources, the genetic engineer must be able to regulate RNA production; that is, the genes must be turned on and off at the appropriate times.

Two mechanisms for regulating messenger RNA production have been extensively studied in bacteria. In the type called **repression**, RNA synthesis is blocked by a particular protein. This process is outlined in the paragraph that follows. A second type of control is called **attenuation**. In attenuation, a short leader of messenger RNA is made, but RNA synthesis is halted before the gene region is reached. Attenuation is described in Appendix I.

Gene control by repressors has been most thoroughly studied with genes that code for enzymes involved in the breakdown of foodstuffs entering bacterial cells. Unless the particular foodstuff is present, there is no point in producing the enzyme that breaks it down. Some genes of this type are turned off by a special protein called a **repressor**, which binds specifically to DNA just in front of the gene it controls (Figure 3-9). As long as the repressor sits on the DNA, RNA polymerase is unable to start producing a message from that gene. If the foodstuff

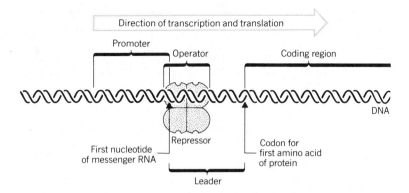

Figure 3-9. **Control of Gene Expression by a Repressor.** *RNA polymerase normally binds to a region of DNA called a promoter. The polymerase then makes a short leader RNA followed by the coding region of the gene. The stretch of DNA that has the information for the RNA leader also serves as a binding site for the repressor. This stretch of DNA is called the operator. When the repressor binds to the operator, RNA polymerase is unable to attach to this region of DNA and thus cannot make the messenger.*

suddenly becomes available and enters the cell, the repressor binds to the foodstuff and leaves its spot on the DNA. As soon as that happens, RNA polymerase binds to the DNA near the beginning of the gene at a region called a **promoter**. The polymerase then promptly makes the messenger for the enzyme required to degrade the foodstuff. The newly-made messenger attaches to ribosomes and is translated, as described earlier in this chapter, producing the degradative enzyme. And as soon as the enzyme is made, it begins to break down the foodstuff. When all the particular foodstuff has been devoured, none is left for the repressor to bind to. The repressor then returns to its spot on the DNA, a region called the **operator**, halting production of the messenger for the degradative enzyme. Thus the cell conserves its energy by producing the specialized degradative enzyme only when that particular enzyme is necessary. An important concept is that each gene controlled in this manner has its own specific repressor protein and contains a region of DNA to which its own repressor specifically binds.

Sometimes several genes are controlled by the same repres-

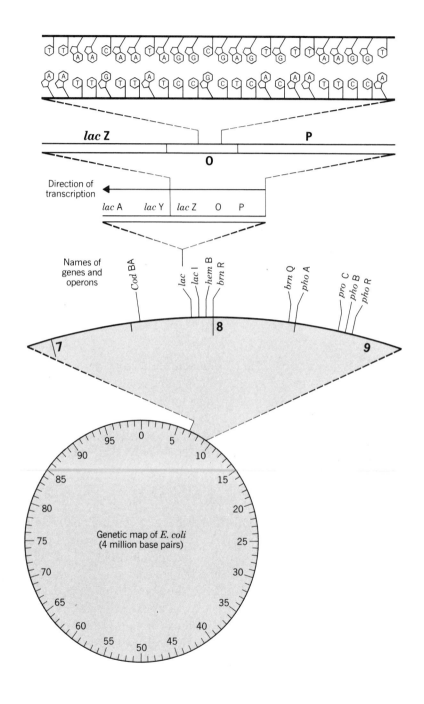

sor; such a unit of genes is called an **operon**. One of the best studied operons is called *lac*, whose three genes are involved in the transport and breakdown of lactose (milk sugar). These genes, *lacZ*, *lacY*, and *lacA*, are arranged in a row on the bacterial chromosome near map position 8 (Figure 3-10). RNA polymerase begins transcribing RNA from the site labeled P in Figure 3-10 and stops after passing through the *lacA* gene. Thus the information from three genes is transcribed into a single RNA molecule. When the *lac* repressor (encoded by the nearby *lacI* gene, Figure 3-10) binds to the *lac* operator (region O in Figure 3-10), transcription is simultaneously blocked for all three genes. The nucleotide sequence is known for much of this region of the chromosome (part of the operator sequence is shown in Figure 3-10), and molecular biologists are now elucidating the detailed chemistry of repressor–operator interactions.

Repression and attenuation are but two regulatory mechanisms that allow cellular machinery to read DNA in a way fundamentally different from the way an editor views motion picture film. The frames in a film must always be viewed one at a time in a linear order—only one scene is accessible to the viewer; all the others are rolled up in the reel. In contrast, many genes are accessible to the cell simultaneously. Thus the cell is able to respond to environmental changes quickly. More complicated organisms like humans, which pass through different

---

Figure 3-10. Organization of An Operon in *E. coli*. *An E. coli DNA molecule is a large circle that, if stretched out, would be 1000 times longer than an E. coli cell. The genes in this DNA molecule are arranged in a circular map divided into 100 units as shown at the bottom of the figure. An enlargement of a region of the genetic map near position 8 shows the location of the* **lac** *operon. The* **lacZ**, **lacY**, *and* **lacA** *genes are involved in the metabolism of lactose (milk sugar). All three genes are transcribed into a single messenger RNA molecule. The promoter (P) is the region where RNA polymerase binds, the operator (0) is the region where the repressor binds, and* **lacZ** *is the gene for the enzyme that degrades lactose. The exact sequence of DNA nucleotides is known for the lactose gene region; the nucleotide sequence where the lactose repressor protein binds to the DNA is shown at the top of the figure.*

developmental stages, may have cells containing a larger number of gene control devices. For example, it might be useful for a cell to inactivate a large block of genes once the genes are no longer needed to help an organism develop. Many molecular biologists are now focusing their efforts on understanding how genes are regulated in humans.

Specialized mechanisms for gene control can raise practical problems for genetic engineers. For example, if an animal gene is placed into a bacterial cell to produce large quantities of a protein such as insulin, a genetic engineer might want to begin with the animal gene turned off so that the bacterial cells will grow normally until the culture contains a large number of cells. Then the engineer might like to turn on the animal gene to produce massive amounts of insulin. But little is known about how animal genes are controlled, and there is no guarantee that the control mechanism would work in bacteria. However, knowing about repressors makes it possible to sidestep the problem by simply splicing the animal gene into bacterial DNA directly behind a bacterial repressor binding site. Thus the bacterial repressor will prevent RNA polymerase from making messenger RNA from the animal gene. The animal gene can then be turned on by adding the correct foodstuff to the bacterial culture; the repressor will bind to the foodstuff and leave the DNA. But now, instead of making messenger for the enzyme that breaks down the foodstuff, RNA polymerase makes the messenger for insulin.

PERSPECTIVE AND REVIEW

In 1953 James Watson and Francis Crick made what may be judged to be the most important scientific finding of our century. They proposed a chemical structure for DNA that guided the thinking of biologists to the point that we can now manipulate the chemistry of heredity. During the past 30 years, much information has been obtained about DNA structure and gene expression. It is probably useful to place some of these concepts into perspective and to point out their relevance to genetic engineering.

1. Living cells are composed of chemicals, many of which are constantly being converted into other chemicals through a series of steps.

2. These chemical conversions are directed by chainlike molecules called proteins, and generally a different type of protein controls each type of conversion. Thus, to alter the characteristics of an organism (i.e., to change the chemical reactions), one must change the proteins controlling these reactions. The protein content of an organism can be changed temporarily by injections (as in the case of insulin) or permanently by genetic engineering.

3. Genetic engineering is based on the observation that the information required to make each protein is stored in another long, chainlike molecule called DNA. This information is arranged into short regions called genes, such that each gene contains the information for the production of one protein. Genetic engineers change a particular protein by changing the information in DNA. Once the change in DNA has occurred, every protein made from the new information is of that new variety. The change is permanent because DNA reproduces itself every time a cell divides.

Gene cloning is a form of genetic engineering. We can review how genes are cloned by adding details to the scheme sketched in Chapter 1. First, DNA is removed from a donor organism and cut with enzymes to make single genes accessible. The DNA fragments are spliced into small, infectious DNA molecules (cloning vehicles), which are used to carry the DNA into microorganisms. The details of cutting and splicing and the biology of cloning vehicles are described in later chapters.

Unfortunately, the cutting enzymes produce thousands of DNA fragments. One of the major tasks of genetic engineers is to find the particular piece of DNA containing information for the protein they wish to change. By putting the DNA fragments into bacteria, it is possible to locate a specific piece of DNA. First, the fragments are transferred into bacterial cells such that each cell receives only one fragment. Each cell multiplies to form a colony; at the same time many copies of the DNA fragments are made. Every colony is then examined for the presence of the DNA fragment being sought. One testing procedure is based on the chemical structures of DNA and RNA and the principle that complementary base pairs tend to stick together. Since messenger RNA from a specific gene is complementary to one of the DNA strands of that gene, it can form a double-stranded structure with DNA. Consequently, messenger RNA can be used to test bacterial colonies for the presence of a specific gene by looking for **RNA:DNA hybrids**. The detection of hybrids is discussed in Chapter 4 and Appendix III.

Once a bacterial colony that contains the gene being sought has been identified, the gene cloner must remove the DNA from the cells in the colony. This is accomplished by purifying the cloning vehicle, which still carries the gene. The gene can then be cut out of the cloning vehicle, spliced behind bacterial regulatory regions, reintroduced into bacterial cells, and cultured for large-scale production of the protein that the gene specifies. Alternatively, the gene can be spliced into a different cloning vehicle to carry it into cultured animal cells, which in turn may be put into whole animals. For the latter scenario to be successful, much more must be discovered about the regulation of genes in animals and about how to get cultured cells to grow in whole animals.

The next chapter continues to describe the biology of DNA, with major emphasis on how DNA reproduces. The important role of complementary base-pairing becomes even more obvious, and the strategies for detecting bacteria carrying cloned genes should become clearer. Chapters 5 and 6 use this background to discuss gene splicing and cloning vehicles.

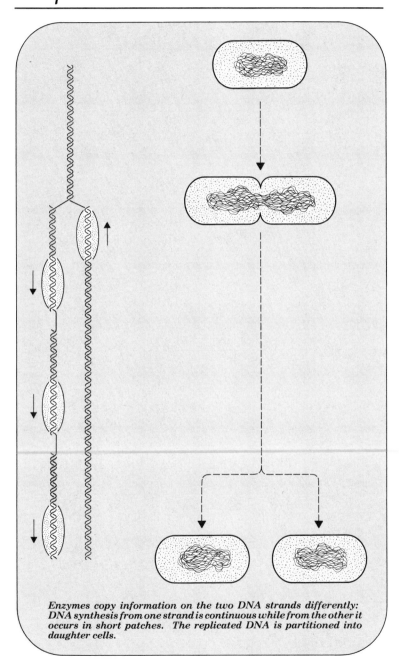

Enzymes copy information on the two DNA strands differently: DNA synthesis from one strand is continuous while from the other it occurs in short patches. The replicated DNA is partitioned into daughter cells.

chapter 4

REPRODUCING DNA

information transfer from one generation to the next

overview

DNA is copied by a group of proteins traveling together along a double-stranded DNA molecule. As the proteins move along the DNA, they separate the two DNA strands and make a new strand adjacent to each old one; one double-stranded molecule becomes two. Each DNA has one old and one new strand, and the nucleotides in these two strands are complementary. Occasionally errors occur while the new strands are being made, and the nucleotide sequence in the new strand is not perfectly complementary to that in the old strand. If not corrected, these errors, called mutations, are passed faithfully from one generation to the next.

The two strands in a DNA molecule are not replicated in exactly the same way. Replication of one old strand occurs by formation of one long, continuous complementary strand. Replication of the other, however, occurs by formation of short patches, which are later connected. The enzyme that connects the patches (DNA ligase) has been

51

obtained in pure form and is now used by gene cloners to splice DNA fragments. Other enzymes involved in DNA replication are used by cloners to make highly radioactive DNA in test tubes. This radioactive DNA is utilized to find bacterial colonies that contain specific cloned genes.

INTRODUCTION

Earlier we defined life as a set of chemical reactions organized in such a way that this set can reproduce itself. The preceding two chapters focused on how DNA and enzymes organize the reactions. Now, to introduce many of the tools used in gene cloning, the aspect of reproduction is examined. For the present discussion, reproduction is defined as the duplication of the information content of the cell followed by segregation of this information into two newly formed daughter cells. Since information is stored in DNA, knowing how DNA duplicates is crucial to understanding reproduction. The next section discusses the machinery responsible for **DNA replication**, the technical term for reproduction of DNA. Understanding **mutations** has been important for deciphering how DNA functions, so the chapter continues with a section on DNA structure and mutation. The third section outlines how enzymes are obtained and handled since enzymes involved in DNA replication are important for manipulating DNA. The final section discusses how radioactive probes are used to locate bacterial colonies containing cloned genes.

DNA REPLICATION

To provide genetic continuity from one generation to the next, DNA not only must be chemically stable, but it also must be copied accurately during replication. If this were not the case, the DNA of an offspring would contain information different from that of its parents. Then the proteins in the offspring would differ from those in the parents, and the characteristics of the two generations would no longer be the same. Accurate copying is accomplished in the following way. The two strands of DNA separate so each can act as a template for formation of a new strand. Nucleotides are aligned along each DNA strand according to the complementary base-pairing rule (Figure 3-3), and they are joined to form a new DNA strand. Two DNA molecules arise from one, and they contain identical information (Figure 4-1). The two DNA molecules move to different parts of the cell, cell division occurs between the DNA molecules, and two daughter cells arise having identical DNAs (Figure 4-2).

Since biochemists can replicate DNA in test tubes by adding a small number of purified components, much is known about the replication machinery. The process can be divided into a number of steps. First, specific proteins form a complex with DNA. Next, the two DNA strands begin to unwind, producing a

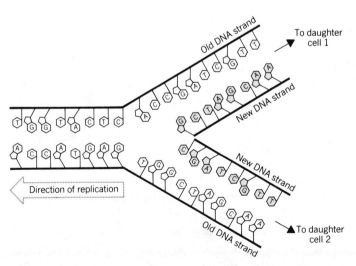

Figure 4-1. Two DNA Molecules Arise from One. *Base-pairing complementarity allows information to be copied exactly. Notice that the set of Information transmitted to the two daughter cells is identical. For clarity, the strands have not been drawn as an interwound helix (see Figure 3-1).*

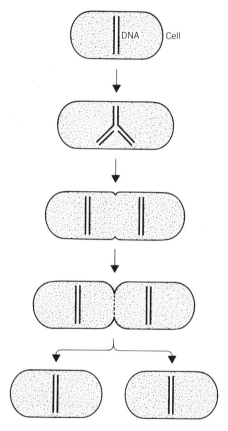

Figure 4-2. Replication and Segregation of DNA. *While the cell is growing larger, its DNA replicates. The two daughter DNA molecules move to opposite parts of the cell. A new cell wall forms between the DNA molecules, and two daughter cells are produced.*

replication fork that moves through the double-stranded DNA molecule as replication occurs. The process is much like unzipping a zipper (Figure 4-1). Thus two single strands are created, exposing the bases. An enzyme complex called **DNA polymerase** binds to *one* of the single strands and moves in the direction of fork movement, closely following the zipper (Figure 4-3). As it moves along the single strand, DNA polymerase mediates

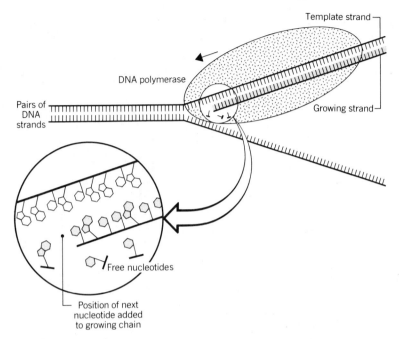

Figure 4-3. Movement of DNA Polymerase. *The DNA strands separate as DNA polymerase and other proteins bind to the DNA. In the replication of one strand, DNA polymerase follows the replication fork, forming a new strand whose nucleotide sequence is complementary to that of the old, template strand. Nucleotides are added to the end of the growing chain one at a time. The order of the nucleotides in the growing chain is determined by the order in the template strand. For clarity, the DNA has been drawn as two parallel strands rather than as a double helix as in Figure 3-1.*

formation of base pairs between free nucleotides (links not yet in a chain) and the linked DNA nucleotides. The alignment obeys the complementary base-pairing rule, so wherever an A occurs in the single DNA strand, a T is aligned opposite it. As soon as a free nucleotide is lined up at the end of the growing chain (Figure 4-3), and opposite to its complement in the template strand, DNA polymerase links it to the new chain. The polymerase moves down the chain one position, aligns the next nucleotide, and links it to the growing chain. Thus a new single-stranded DNA chain is formed using the information contained in the old one. Of course as the new chain forms, it is already base-paired with the old one; a double-stranded DNA molecule containing one new strand and one old one is produced.

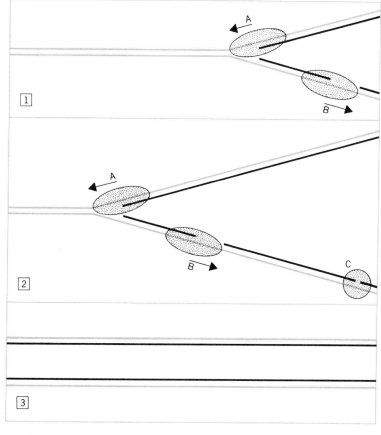

Figure 4-4. Discontinuous DNA Synthesis. *(1) One DNA polymerase complex (A) moves continuously along an old single strand in the direction of replication fork movement while the other (B) moves in the opposite direction. (2) Polymerase B synthesizes short patches of DNA. Small gaps are filled in by another type of DNA polymerase, and the DNA fragments are spliced together by DNA ligase (C). (3) The result is two double-stranded DNA molecules.*

A particularly interesting detail that fits the scheme just described is that DNA polymerase always travels in the same direction along a DNA chàin. Like a motion picture film, DNA has directionality. The nucleotides are bound together much like a chain of elephants hooked trunk to tail, and the polymerase recognizes this directionality. This detail becomes interesting when you discover that the two strands in double-stranded DNA run in opposite directions (see Figure 3-2). As the replication fork moves through the DNA molecule, unzipping the DNA and making single strands available for replication, DNA polymerase on one strand will follow the fork (Figure 4-4), moving in the same direction as the fork. Polymerase on the other strand, however, must move in the opposite direction, away from the fork. Synthesis *opposite* to fork movement occurs in short patches, which in bacteria are about 1000 nucleotides long (see frontpiece, Chapter 4). Once the polymerase has made one patch, we could imagine that it leaps back toward the vee in the fork and begins making another piece of DNA until it runs into the patch of new DNA it had just laid down. Then the polymerase must again catch up with the moving fork. Whether the polymerase actually leaps toward the moving fork is not known; molecular biologists are currently testing ideas that don't require the polymerase to leap. It is clear, however, that DNA polymerase is unable to join the two patches of new DNA together. Another enzyme, called **DNA ligase**, is required to connect the patches. Ligase is an important tool genetic engineers use to splice DNA fragments together.

DNA STRUCTURE AND MUTATIONS

Occasionally errors are made during DNA replication, and the errors may be passed to the next generation of cells. Changes in genetic information can have serious consequences. Consider the case outlined in Figure 4-5, in which a gene is copied incorrectly. In an offspring a single nucleotide pair might be

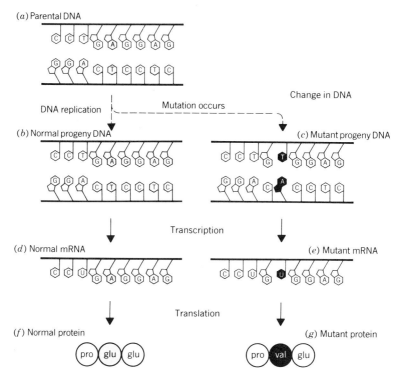

Figure 4-5. A Point Mutation Changes the Sequence of Amino Acids in a Protein. *DNA replication is very accurate, so the nucleotide sequence in the parental DNA (a) is identical to that of normal progeny DNA (b). Occasionally an error is made. In the example shown, a particular AT base pair, in parental DNA (a) changed to a TA pair in the mutant progeny DNA (c). This results in a conversion of a particular G-A-G codon in normal messenger RNA (d) into a G-U-G codon in mutant messenger RNA (e) (remember that RNA contains U instead of T). In the resulting proteins, G-A-G codes for the amino acid glutamic acid (glu) (f) while G-U-G codes for valine (val) (g). The two amino acids have very different chemical properties. Since the structure of the resulting protein is determined by the precise order of the amino acids, the mutant protein will differ significantly from the normal protein. The differences between the normal and mutant molecules shown are identical to those found between healthy people and patients suffering from sickle-cell anemia.*

changed in the gene; for example, an AT pair might be convert-ed to a GC, CG, or TA pair. RNA polymerase is unable to tell that an error has been made, and it faithfully transcribes the misinformation into messenger RNA. Thus in the offspring the messenger RNA from the gene will always contain the error. Likewise, the translation machinery, which converts the infor-mation in the messenger RNA into protein, is not able to distinguish between correct and erroneous messengers. Conse-quently, an incorrect amino acid may be inserted into the protein specified by the gene containing the error. An incorrect amino acid will often change the structure of the protein enough to make it inactive. Since inactive proteins frequently alter the chemistry of cells and organisms, it would not be surprising to find that the offspring looked different from the parent or acted in different ways. Such an altered organism is called a **mutant**, and if the mutation (i.e., the nucleotide change in the DNA) has occurred within an essential gene and produced an inactive protein, the mutant organism will die. Occasionally mutation produces a better protein in the offspring, making it better able to survive and reproduce in that particular environment. In such cases the mutants may eventually become the dominant type of organism in the population, and a small step in evolution will have occurred.

A number of different types of mutation have been discov-ered (Figure 4-6). In addition to the example given above, changing a single nucleotide occasionally converts a normal triplet codon into a stop codon—then the translation machinery prematurely stops making the protein (Figure 4-6c). If such a mutation occurred near the beginning of the gene, only a small fragment of the protein could be made, with obviously serious effects. In another kind of mutation a nucleotide is lost during replication; that is, the replication machinery skips a letter (Figure 4-6d). In this case the messenger RNA is thrown out of the correct reading frame, and many incorrect amino acids are placed in the protein. This is called a frameshift mutation. In still other cases, a large chunk of DNA is lost (Figure 4-6e). Fortu-

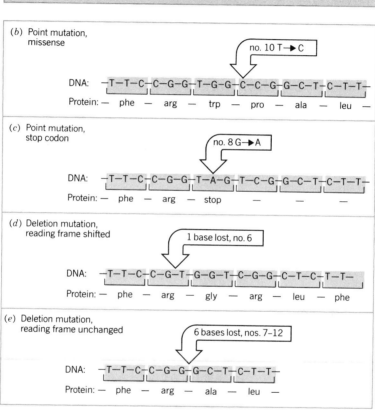

Figure 4-6. Common Types of Mutation. *(a) A normal nucleotide sequence for one strand for DNA codes for a protein having the six amino acids listed (phe, arg, trp, etc). The codons for the amino acids are indicated by brackets in the DNA above the respective amino acids. (b) If a T is changed to a C (arrow), the resulting mutant protein has a proline where serine is normally located. This change does not affect the joining of amino acids and is called a **missense** mutation. (c) If a G is changed to an A as in the example shown (arrow), a nonsense (stop) codon is created and protein synthesis halts. The mutant protein is shorter than the normal protein. (d) Deletion of one base*

nately, mutations rarely occur in the absence of chemical agents and as a result most species have stable characteristics.

The accuracy of the replication machinery is very impressive. In bacteria, detectable mutations (errors) in a given gene arise at a frequency of only one in a million cells per generation, even though the replication apparatus copies DNA at a rate of 50,000 base pairs per minute. The key to understanding this process lies in the principle of base-pairing complementarity. The important point is that biological molecules recognize each other by fitting together like locks and keys. Nearly perfect fits are required for two molecules to bind together, and when the fit is good, the binding can be strong. Thus, as long as the rule illustrated in Figure 3-3 is obeyed (namely, that A bind only to T and G only to C), there will be no errors and replication will proceed properly.

Within this framework, any chemical that converts one letter to another in the old DNA strand is capable of "tricking" the replication machinery into inserting the wrong nucleotide into the new DNA strand as it is made. Chemicals that alter the information in DNA are called **mutagens.** Some mutagens are thought to be involved in certain kinds of cancer.

HANDLING ENZYMES

Thus far enzymes have been considered to be tiny agents that somehow join nucleotides to form RNA or DNA chains, join amino acids to form protein chains, and in general direct the chemical reactions of the cell. To explain how enzymes are used as tools for genetic engineering, it is necessary to describe some

←——————————————————————

throws the reading frame out of register, and incorrect amino acids (gly, arg, leu) occur in the mutant protein. (e) Removal of six bases produces a deletion mutation and a protein missing two internal amino acids (trp, ser).

of the technical aspects of handling enzymes. Methods for obtaining DNA polymerase are outlined below to illustrate how enzymes are detected and purified.

To be useful for gene cloning, enzymes must be removed from living cells. The first step of purification is to obtain a large batch of cells, about a pound of cells in the case of bacteria. Then the cell walls must be broken to liberate the enzymes. This can be done in a number of ways; one method is to combine the cells with tiny glass beads and grind the mixture in a blender. The mixture of cellular components is called a **cell extract**. Then the enzyme of interest is separated from other cell components by a series of physical manipulations. Separation is possible because every type of enzyme is physically and chemically different from every other type of molecule in the cell. For example, DNA polymerase is relatively small compared to many cell components. Thus one purification step might be to allow the large cellular debris to settle to the bottom of a test tube (Figure 4-7). DNA polymerase is expected to remain in the fluid at the top of the tube. A crucial part of this process is determining whether DNA polymerase has settled out. An **assay** for the enzyme must be developed, that is, a way to detect and follow the enzyme through a series of purification steps.

To develop an assay for a particular enzyme, a biochemist must make some guesses about the chemical reaction stimulated by the enzyme in living cells. From the discussion in the first part of this chapter, one could imagine that DNA polymerase, using a single-stranded DNA template, causes free nucleotides to link together to form a DNA chain. Figure 4-8 shows how this concept can be used to determine whether an extract of broken cells contains DNA polymerase. Generally the number of free nucleotides that are linked together in a test tube is very small, and it is necessary to have a very sensitive measure for this conversion. Radioactive isotopes provide the key to making such measurements because very small amounts can be detected.

Operationally, one type of radioactive nucleotide (e.g.,

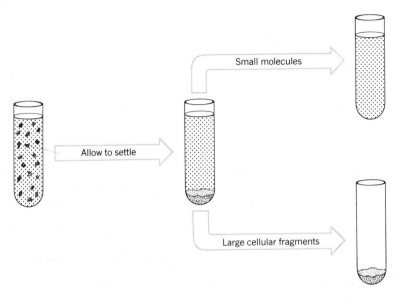

Figure 4-7. Fractionation of a Cell Extract. *The force of gravity can be used to separate molecules on the basis size. After the large molecules have been allowed to settle to the bottom of the tube, the upper solution can be carefully removed with a pipette and placed in a second tube.*

radioactive T) is mixed with the other three nonradioactive nucleotides, single-stranded DNA template, and the cell extract that contains the DNA polymerase. After incubating the mixture for an hour or so at 37° C (body temperature), cold acid is added to the mixture. DNA molecules clump together to form a white, stringy **precipitate**, a solid that settles to the bottom of the test tube. If the acid-containing mixture is then poured into a funnel lined with filter paper, the solution will pass through the filter paper, but precipitated DNA will stick to it. Free nucleotides are relatively small, and they do not precipitate when acid is added to the mixture; consequently, they do not stick to the filter paper. The only radioactive molecules that can stick to the filter paper are the nucleotides that have become a part of the DNA. Thus, the amount of radioactivity on the filter paper is a

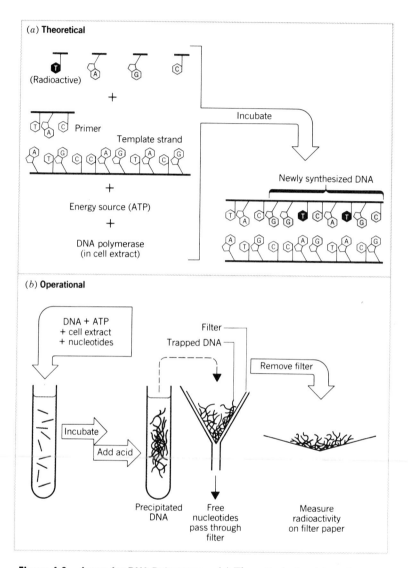

Figure 4-8. Assay for DNA Polymerase. (a) Theoretical. A mixture is prepared that contains the four nucleotides [A, T (radioactive), C, and G], an energy source (**ATP**), single-stranded DNA template containing a short double-stranded region that provides a **primer** (DNA polymerase always requires a primer to begin its action; it does not work on single-stranded DNA alone), and a sample containing DNA polymerase. During incubation

measure of the number of nucleotides that have been converted into DNA, which in turn is a measure of the DNA polymerase activity present.

Molecular biologists use many different enzymes. For each one, a different assay must be devised to follow the enzyme during purification. However, the principle is the same for each assay: there must be a way to distinguish between the reaction **substrate** (the molecules you start with) and the reaction **product** (the molecules you end up with). The molecular biologist measures the speed at which substrate is converted into product. In the example given above, free nucleotides are the substrate of DNA polymerase, and nucleotides incorporated into DNA are the product of the reaction. The speed of the reaction is determined by measuring the amount of radioactivity sticking to the filter paper for each minute the total reaction mixture is incubated at 37° C. The speed of the reaction indicates the amount of enzyme present in the cell extract. Thus, after separating cellular components into different test tubes (Figure 4-7), the molecular biologist uses the assay (Figure 4-8) to determine which tube contains the enzyme.

A number of methods are used to separate subcellular components. In the procedure called column chromatography a glass tube is mounted vertically and filled with one of many different solid materials (e.g., ground-up paper that has been chemically treated). The cell extract is allowed to slowly flow

←——————————————————————————————

DNA polymerase joins the free nucleotides together to form DNA. The newly made DNA will be radioactive because of incorporation of radioactive T. (b) Operational. The mixture described in a is added to a test tube and incubated to allow the reaction to occur. Acid is added to stop the reaction and to cause the DNA molecules to clump together. The mixture is poured through a piece of filter paper, and the clumped DNA is trapped on the paper. If any radioactive DNA has been made in the reaction, it will stick to the filter paper because of its large size. In contrast, molecules of T (radioactive) that not were incorporated into DNA are washed through the filter because they are small. The radioactivity on the filter paper is measured with an instrument called a liquid scintillation counter.

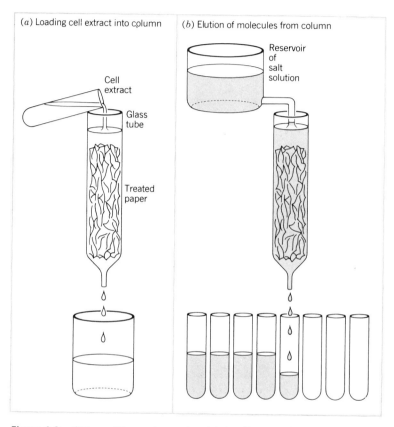

Figure 4-9. Column Chromatography. *(a) A cell extract is passed through a solid matrix (often a chemically-treated material similar to finely ground paper) to which protein molecules stick with varying degrees of tightness. (b) The proteins can be removed from the column by passing a dilute salt solution through the column. Because of size, some molecules will come through sooner than others. All of the salt solution is collected in a series of test tubes, and each tube is assayed to determine which one contains the protein being sought.*

through the paper packed in the glass tube, eventually dripping out the bottom of the tube. Some molecules bind tightly to the paper, others loosely. Thus, when the molecular biologist passes a dilute salt solution through the paper, some molecules come out after very little salt solution has passed through, whereas others require extensive washing. By collecting the salt solution in a series of test tubes (Figure 4-9), the molecular biologist can separate various molecules from each other. Then each tube can be assayed to determine which one contains the enzyme being sought. By combining several chromatographic procedures like the one described above, it is possible to separate an enzyme from all other cellular components.

PROBES TO FIND CLONED GENES

The concepts developed in the discussions of gene expression and DNA replication can now be used to add important details to the general strategy for cloning genes. DNA is first cut in specific places with a purified enzyme called a **restriction endonuclease**. As mentioned before, many DNA pieces are produced, and they are spliced into cloning vehicles using another purified enzyme, DNA ligase. Both enzymes have been purified by strategies similar to those described in the preceding section, and both are commercially available. The cloning vehicles carry the DNA fragments into E. coli cells, which are subsequently grown into separate colonies on agar plates. At that stage the gene cloner breaks open the cells and converts all the double-stranded DNA in each cell into single-stranded molecules, thus making the bases available for base-pairing. The final step is to treat the broken E. coli cells with a radioactive **probe** (DNA or RNA known to be complementary to the gene one wishes to isolate). The probe will form base pairs only with DNA from the particular bacterial cells that contain the gene being sought. This process of forming base pairs between two different **nucleic acid** molecules is called nucleic acid hybridization (Figure 4-10). The bacterial colony containing those cells will then become radioactive and can be identified as described in Appendix III. Colonies lacking the cloned gene will not become radioactive.

Radioactive probes can be obtained in several ways. In some

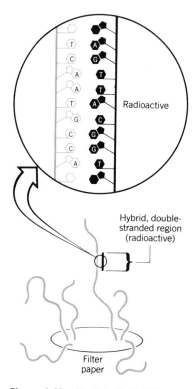

Radioactive

Hybrid, double-
stranded region
(radioactive)

Filter
paper

Figure 4-10.   Nucleic Acid Hybridization. *Under the appropriate conditions two complementary single-stranded nucleic acids will spontaneously form base pairs and become double-stranded. If single-stranded, nonradioactive DNA (shaded) is fixed tightly to a filter and then incubated in a solution containing single-stranded, radioactive DNA (solid), double-stranded regions will form where the two types of DNA have complementary nucleotide sequences; the radioactive DNA will become indirectly bound to the filter through its attachment to a specific region of nonradioactive DNA (open). By measuring the amount of radioactivity bound to the filter, one can estimate the relatedness between two DNAs.*

cases it is possible to isolate messenger RNA from the gene one wishes to clone. Radioactive messenger RNA to be used as a probe can be obtained by growing the cells in the presence of radioactive nucleotides. However, it is often difficult to obtain natural messenger RNA containing enough radioactivity to test the bacterial colonies for cloned genes because so much radioactivity must be added to the cells that they tend to die. Consequently, the messenger RNA is usually employed as a template for the synthesis of DNA using **reverse transcriptase**, an enzyme purified from **RNA tumor viruses**. Since the DNA product, called **complementary DNA** (cDNA), is formed in test tubes, it can be made highly radioactive by using radioactive nucleotides to form the DNA.

A more general way to obtain a probe involves first purifying the protein product of the gene being sought. The order of the amino acids in the protein can then be determined by chemical analysis. Since every amino acid corresponds to a triplet of nucleotides in the DNA of the gene, it is possible to predict the nucleotide sequence of the gene from the sequence of amino acids (the prediction is not exact because most amino acids are encoded by several different codons). The next step is to chemically synthesize a short stretch of DNA, one base at a time, that would have a sequence identical to that predicted for the gene. Instruments are available to determine the amino acid sequence of the protein and to synthesize the short DNA piece. A purified enzyme called a kinase is used to add radioactive phosphorus to the synthetic DNA fragment, making it a highly radioactive probe.

PERSPECTIVE

By the early 1950s the importance of DNA to heredity was widely recognized, and when Watson and Crick realized how the two DNA strands are arranged, biochemists set out to determine how DNA replication works. Enzyme assays and protein purification procedures were developed, and by the early 1970s a number of proteins that participate in DNA replication had been purified. Consequently, when methods were found to cut DNA into specific pieces, replication proteins that would splice the pieces together were already available. In addition, other replication proteins were known that could synthesize highly radioactive DNA to be used as probes for locating bacterial colonies containing cloned genes. Thus information from a number of lines of research was used to develop gene cloning strategies. These strategies are now being used to study aspects of DNA replication that we still don't understand, such as how chromosome replication begins and ends.

chapter 5

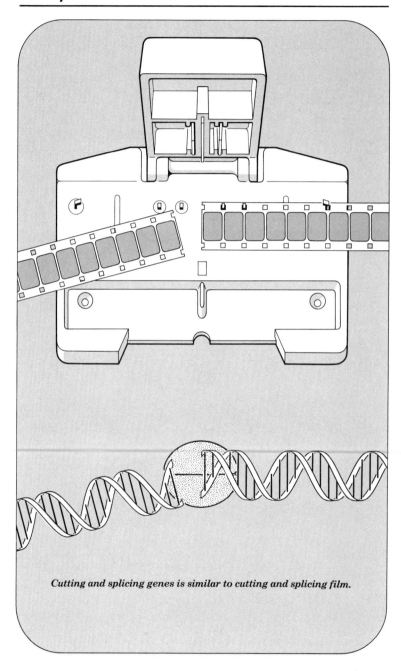

Cutting and splicing genes is similar to cutting and splicing film.

chapter 5

GENE
SPLICING

*cutting and rearranging
informational molecules in
the laboratory*

overview

*Gene cloners move specific bits of genetic information
from one DNA molecule to another by cutting and splicing procedures that utilize specific enzymes. DNA is very
long and contains a large number of sites at which cutting
can occur. Consequently, cutting and splicing often results
in many different combinations of fragments being joined.
Biochemical methods based on the principle of complementary base-pairing are used to detect and locate a specific
combination of spliced fragments.*

The enzymes that cut DNA are called restriction endonucleases *and they recognize and cut at specific DNA
sequences. Consequently, cutting DNA with these enzymes produces DNA fragments having discrete lengths.
Enzymes can be obtained that recognize different sites in a
DNA molecule, making it possible to cut DNA at a wide
variety of locations. Cuts by different enzyme types produce*

73

DNA fragments having different sizes; two enzymes cutting the same DNA at the same time will cut in different places and will produce smaller fragments than will either enzyme cutting alone. It is possible to locate the cutting sites of one enzyme relative to another by comparing the sizes of the DNA fragments produced by two enzymes cutting simultaneously with the sizes of DNA fragments produced by each enzyme cutting alone. This type of analysis produces a restriction map *characteristic of the particular DNA being studied.*

A mutation occurring in a DNA molecule at a site where a restriction endonuclease normally would cut will often prevent the enzyme from cutting at that point; consequently, a larger DNA fragment will be produced when mutant DNA is cut. It is now possible to detect some genetic diseases in human fetuses by analyzing the sizes of the DNA fragments produced by the cutting enzymes.

INTRODUCTION

Since the gross aspects of information organization in DNA are easily described by means of analogies between DNA and motion picture film, film metaphors are used to begin describing gene cutting and splicing. Imagine a film editor who feels that he can make an exciting, profitable, new motion picture by clipping out a scene from a John Wayne film and sticking it into a short Mickey Mouse cartoon. The spliced film will be longer than the original Mickey Mouse cartoon and, depending on which John Wayne scene was moved and where it was placed, the resulting movie will make more or less sense. A new motion picture will be created. In the same sense, gene splicing creates new forms of life.

Like the film editor, a gene cloner needs two tools: scissors and splicing tape. Biologists use specific enzymes as scissors. These enzymes obey an important specificity rule: they cut only at specific places in the DNA. The specificity arises because the cutting enzymes recognize certain short sequences of nucleotides in the DNA. If this rule were applied to the task of a motion picture editor, he would be able to cut only where specific events occur in the movie. For example, the editor might have to follow a rule allowing him to cut the film ONLY where someone shoots a gun. EVERY time anyone shoots a gun in ANY movie, the editor MUST cut. If nobody ever shoots a gun in a particular movie, the editor would not be allowed to cut that film; consequently, he would be unable to carry out splicing steps. Obviously, there may be many gun shot episodes in a given movie. Such a film will be cut into many pieces. Notice that the specificity rule leaves the editor with little control over his scissors; they cut automatically wherever they see a particular sequence of events. Consequently, before an editor can begin splicing, he must find the desired scene by sorting through the many John Wayne film fragments created by his scissors. Alternatively, the editor could splice each John Wayne fragment into a separate copy of the Mickey Mouse cartoon, creating a large number of new movies; he would then have to view each new movie to find the desired one. With film, the latter method is inefficient because an editor would have to make many nonproductive splices. He would then have to spend time viewing the many new films to find the desirable one. Gene cloners, however, use the second strategy because they can easily "look" at thousands of splicing events and find new arrangements of nucleotide sequences.

RESTRICTION ENDONUCLEASES

Restriction endonucleases are a group of enzymes that correspond to the scissors in the analogies developed above. They recognize specific nucleotide sequences in DNA, often four or

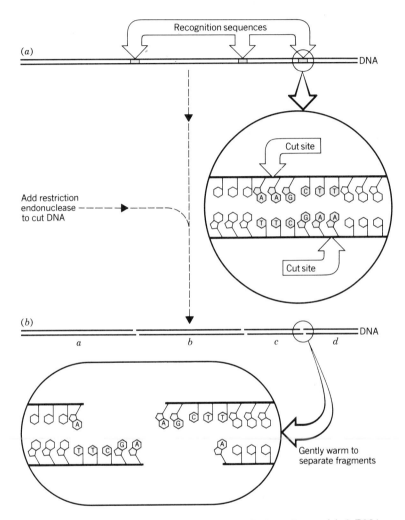

Figure 5-1. Cleavage of DNA by a Restriction Endonuclease. *(a) A DNA molecule, depicted as two parallel lines, may contain many short nucleotide sequences recognized by restriction endonucleases. (b) When a restriction endonuclease is added to the DNA, it binds to the DNA and cuts it. Some of these enzymes produce staggered cuts. The DNA molecule in the example is converted into four shorter molecules, a, b, c, and d, each with "sticky ends" that can form base pairs with each other.*

six base pairs long, and cut both strands of DNA within the **recognition site**. In some cases the two DNA strands are not cut opposite each other—rather, the cuts are staggered. In the case illustrated in Figure 5-1 the cuts are offset by four nucleotides. Once the cuts in this example have been made, only four base pairs are holding the two DNA strands together. When the cut DNA is gently warmed, the four base pairs will break apart, and the DNA molecule will separate into fragments.

Many restriction endonucleases are commercially available, and most differ in their recognition sequences. This collection of enzymes gives molecular biologists many different cutting options.

LIGATION

Some restriction endonucleases generate staggered cuts (Figure 5-1b) that molecular biologists call **sticky ends**. The four nucleotides in the single-stranded ends of the DNA molecules are complementary to the ends of other molecules generated by cutting with the same restriction endonuclease. Thus, when two DNA molecules having complementary ends collide, the single-stranded ends form base pairs, and the two molecules tend to stick together (Figure 5-2). Chapter 4 mentioned DNA ligase, an enzyme that performs the essential function of joining DNA molecules together after DNA replication (Figure 4-4). If this enzyme is present when two DNA molecules having sticky ends come together, it will repair the breaks that had been introduced by the restriction endonuclease. Thus splicing can be accomplished by simply mixing together DNA molecules having complementary sticky ends and adding DNA ligase. The technology is now so advanced that sticky ends are not required; a type of DNA ligase has been discovered that will join blunt ends. Such an enzyme is useful to join ends created by restriction endonucleases that do not produce sticky ends.

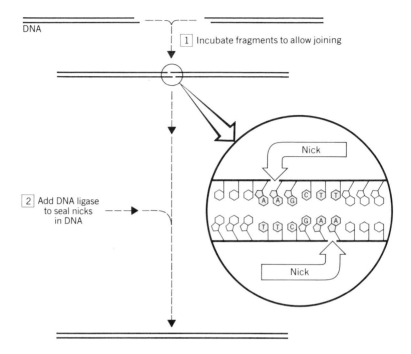

Figure 5-2. Splicing Two DNA Fragments Together. *(1) Two DNA molecules with complementary sticky ends are mixed and incubated. The molecules collide, and base pairs form (2) The strand interruptions (nicks) are enzymatically sealed by DNA ligase.*

CUTTING AND SPLICING

The motion picture metaphor is useful when considering the details of cutting and splicing. First, imagine a large vat (representing a test tube) into which you have placed many, many unrolled copies of a particular John Wayne movie and a particular Mickey Mouse cartoon. The two films might represent many copies of an animal DNA and a small, infectious DNA (the cloning vehicle), respectively, in solution in a test tube. Now imagine that you and some friends enter the vat with splicing scissors and cut EACH film wherever anyone is shown firing a gun. This step corresponds to the addition of many identical restriction endonuclease molecules to the test tube to cut the DNAs. Soon, many film fragments fill the vat. You and your friends leave the vat, so no more cutting can occur. The mixture is then stirred and the fragments collide. In the case of DNA in a test tube, some of the DNA collisions result in the ends sticking together. If DNA ligase is added, the fragments will become permanently spliced.

As mentioned at the beginning of the chapter, the goal of gene splicing is analogous to splicing one specific scene from the John Wayne movie into a COMPLETE Mickey Mouse cartoon. This chore can be simplified in two ways. First, before any cutting is done, tape the ends of the cartoon together to form a circle. Second, choose a cartoon containing only one gunshot sequence, so that only one cut will occur in the cartoon. The single cut in the cartoon will produce sticky ends, which will attach to the ends of a John Wayne fragment. By starting with a circular cartoon, which can be cut only once, you ensure that the cartoon will always remain intact: no fragments will be created that must reattach to produce a complete cartoon. Eventually a John Wayne fragment will collide with a sticky end of the cartoon, and the two can be spliced together. The remaining sticky end of the cartoon will at some time collide with the free end of the John Wayne fragment already spliced onto the other end of the cartoon. When this second splicing event occurs, a circle will have been formed, containing a complete cartoon plus a John Wayne fragment. If the two taped-together ends are released, the film can be projected from beginning to end. The new film will make sense except for a brief interruption where the John Wayne sequence has been inserted. This probably would not interrupt the cartoon any more than would a television commercial. The general scheme for DNA is outlined in Figure 5-3.

In the vat, collisions between ends occur randomly, and much of the time the two ends of the cartoon simply rejoin. Likewise, John Wayne fragments attach to each other. In general, quite a mess is created. Occasionally, however, a single John Wayne fragment will collide with the cartoon. But since

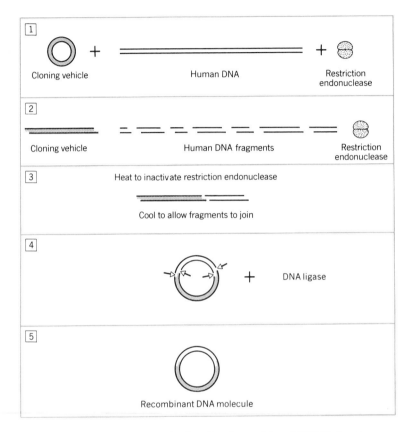

Figure 5-3. General Scheme for Forming Recombinant DNA Molecules.
(1) Circular cloning vehicle DNA, human DNA, and a restriction endonuclease are mixed. (2) Both DNAs are cut, producing a linear cloning vehicle and many human DNA fragments. All DNAs in the mixture have complementary, sticky ends. (3) Occasionally a human DNA fragment will attach to one end of the cloning vehicle. (4) Eventually both ends of the human DNA fragment will attach to the respective, corresponding ends of the cloning vehicle. When DNA ligase is added, the discontinuities in the DNA strands (arrows in 4), are sealed, producing a circular recombinant DNA molecule with no breaks in the DNA strands (5).

there are thousands of different John Wayne fragments, only rarely will any PARTICULAR fragment attach to the cartoon. Thus, in genetic engineering, trillions of DNA molecules may have to be incubated together before there is a reasonable chance that the desired fragment will attach to the cloning vehicle.

RESTRICTION MAPS

Restriction endonucleases have been extracted from a large number of bacteria. These enzymes appear to be part of the natural defense mechanism protecting bacterial cells against invasion by foreign DNA molecules such as those contained in viruses (described in Chapter 6). A crucial element to this protective device is that the **nuclease** must discriminate between the cell DNA and the invading DNA; otherwise the cell would destroy its own DNA. The recognition process involves two elements. First there are specific base sequences that act as targets for the nuclease. Second, the cell is able to place a protective chemical signal on all the target sequences that happen to occur in its own DNA. The signal modifies the DNA and prevents the nuclease from cutting. Invading DNAs would lack the protective signal and would be chopped up by the nuclease unless the invaders came from cells that had put the correct protective signals on them. It turns out that restriction endonucleases from different organisms recognize different target sequences. Thus restriction endonucleases purified from different organisms become enzymatic tools that can be used to cut at different sites in DNA.

In addition to their role in DNA splicing procedures, restriction endonucleases play an important part in the analysis of nucleotide sequences of DNA molecules (described in Chapter 8). The initial step in these analyses is to construct a **restriction map**; a procedure is outlined below to provide familiarity with these cutting enzymes. First, return to the analogy between DNA and motion picture film, and imagine that you wish to study a film 1000 feet long. You are not allowed to use a movie projector, and the film has no ends: it is a circle. You could begin to examine this tangled mass if you could cut the film at SPECIFIC places to produce DISCRETE, manageable pieces. Now suppose that you have three kinds of scissors: type A, which cuts when a gun is fired, type B, which cuts when a dog bites a man, and type C, which cuts whenever a car door is opened (these scissors correspond to different restriction endonucleases cutting at specific sites on DNA). In the film you are studying, assume that a gun is shot only once. Therefore type A scissors will cut only once, producing two ends and giving you a reference point:

If in your film men are bitten by dogs three times, the circle will be cut into three fragments by type B scissors. You can measure the lengths. Suppose that the fragments are 100, 300, and 600 feet long.

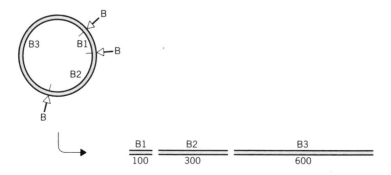

Likewise if a car door opens twice in the film, type C scissors will cut two times, producing two fragments. Suppose that these fragments are 200 and 800 feet long.

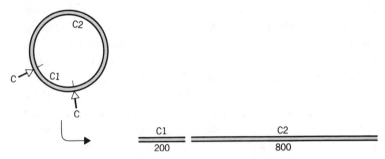

Even smaller fragments can be obtained by cutting with more than one scissors type. For example, type A and type B combined should produce four fragments. Suppose that you find the resulting lengths to be 50, 100, 300, and 550 feet. In such

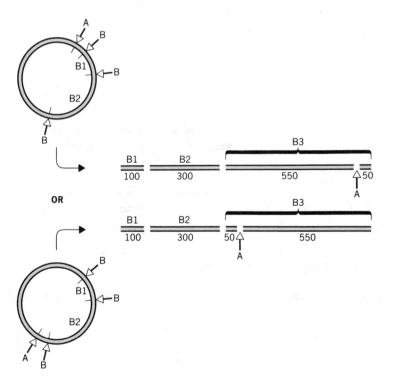

a case the type A cut must have occurred within the 600-foot fragment created by the type B enzyme (B3), producing two new fragments (50 and 550). There are only two ways the film can be arranged to produce this result, and in both cases the gunshot (A) occurs between the two most widely spaced dog bites (B). Thus the map is beginning to develop. At this point you cannot tell whether cut A is nearer to fragment B1 or to fragment B2.

The next combination involves cutting with type A and type C scissors. This time you find three fragments, which are 200, 375, and 425 feet long. The type A cut must have occurred within the 800-foot C2 fragment, generating two new pieces 375 and 425 feet long. Again there are two places where the A cut could be relative to the two C cuts.

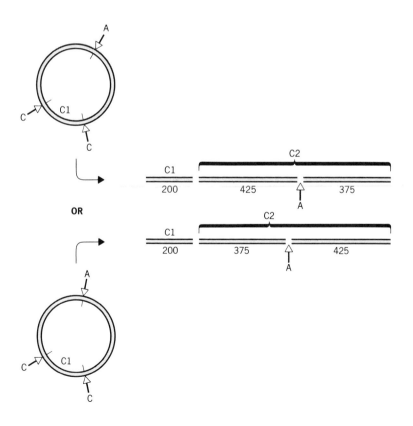

You can also cut with a combination of type B and type C enzymes. Suppose this produces five pieces, 75, 100, 125, 225, and 475 feet long. The 100-foot piece is B1, but B2 (300 feet) and B3 (600 feet) have disappeared. This must mean that one C-type cut occurs in B2 and the other in B3. The 75- and 225-foot fragments add up to 300 feet, which corresponds to B2, so one C-type cut is 75 feet from one end of B2. The sum of 125 and 475 is 600, the value of B3; thus the other C-type cut is 125 feet from one end of B3. Since the two C-type cuts are either 200 or 800 feet apart on the accompanying circular map, there is only one way the map can fit together.

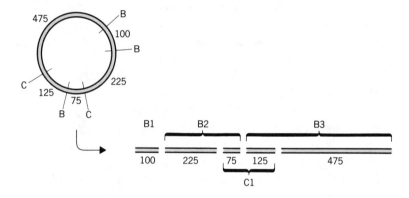

By adding the results from the A + C, the A + B, and the B + C combinations, it is possible to determine the position of the A cut. Since A is in B3, only 50 feet from a B cut, and far from the C cuts, A must map as follows.

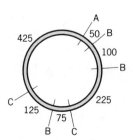

When all three types of cut are made simultaneously, six fragments will be produced.

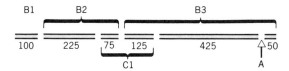

At this stage in the film analogy one would say that the sequence of events is gun shot, two dog bites, a car door opening, another dog bite, and finally another car door opening. The film distance between the events is also known.

In the case of DNA, the size of each of the fragments is measured by a technique called **gel electrophoresis**. A semisolid material like gelatin or agar is first molded into a slab. A small well is formed in the gel, and a mixture of DNA fragments is placed in the well. An electric field is then applied across the gel, forcing the DNA molecules to move through the gel. DNA molecules having the same size move together as a group, but smaller DNAs move faster through the gel than larger ones. If trillions of identical DNA fragments are present, you can see them as a band after staining the gel. Fragments of different lengths produce different bands (Figure 5-4). The size of the DNA in each band is determined by comparing how far a given band moved into the gel relative to bands of DNA molecules whose lengths are already known. If enough identical DNA fragments are present in a band, biochemical methods described in Chapter 8 can be used to determine the nucleotide sequence of the DNA fragment. By knowing the map order of the fragments (using the type of logic outlined in the preceding paragraph), it is possible to fit together the short nucleotide sequences from all the fragments like a jigsaw puzzle, eventually obtaining the nucleotide sequence of the entire DNA molecule.

Restriction mapping has also become a diagnostic tool for prenatal detection of certain genetic diseases. In sickle-cell

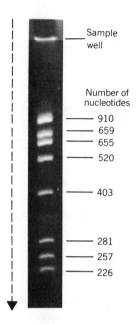

Sample well

Number of nucleotides

—— 910
—— 659
—— 655

—— 520

—— 403

—— 281

—— 257

—— 226

Figure 5-4. Display of Restriction Fragments of DNA by Gel Electrophoresis. *A mixture of DNA fragments was placed in a sample well (slot) cut in a gel. The DNA was driven into the gel by an electric field. Smaller fragments moved faster than larger ones, so when the electric field was turned off, the fragments had separated. The fragments were stained, and the gel was photographed. Each band represents many DNA molecules having the same length. The direction of DNA movement is from top to bottom; the number of nucleotide-pairs in each fragment is indicated at right in each fragment. Photograph courtesy of Richard Archer, Lasse Lindahl, and Janice Zengel, University of Rochester.*

anemia, a single base change is responsible for the disease (see Figure 4-5). The base change occurs in a restriction endonuclease recognition site, and the site is so altered by the change that it is no longer recognized by the enzyme. Thus, the enzyme does not cleave at that location. Consequently, the restriction map of DNA from a sickle-cell fetus differs from that obtained from normal DNA. The principle is illustrated in Figure 5-5 using the restriction mapping example given earlier in this

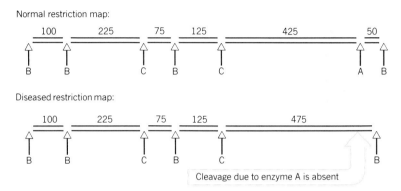

Normal restriction map:

Diseased restriction map:

Cleavage due to enzyme A is absent

Figure 5-5. Diagnosing a Genetic Disease by Restriction Mapping. *A hypothetical example is shown where a change in the nucleotide sequence in DNA causes a disease and also eliminates a known restriction site (site for enzyme A). Thus DNA from a diseased cell has an altered restriction map. Cells from higher organisms such as humans have two copies of each DNA molecule, one derived from the mother and one from the father. Although these molecules are never identical over their whole length, they may be identical in the region being examined. If in this region DNA molecules derived from the parents are both normal or both pathogenic, the analysis produces the results shown in these two diagrams. If, however, the DNA from one parent is normal but the DNA from the other parent is pathogenic, a mixture of the restriction fragments is observed. Sickle-cell anemia is known to be due to a single nucleotide change in DNA (see Figure 4-5), and it is now possible to diagnose the disease by examining restriction maps of fetal DNA.*

chapter. In this example, the base change causing the disease occurs in the recognition site for the type A restriction endonuclease, eliminating that site from the DNA. The map from the diseased fetus would then be easily distinguishable from that of a normal fetus: two normal fragments (425 and 50) would be replaced by a new fragment (475) in the mutant DNA.

PERSPECTIVE

DNA molecules are so long and contain so much information that until the discovery of restriction endonucleases there seemed to be little hope of determining extensive nucleotide sequences. Now that DNA molecules can be cut into discrete, manageable fragments, determining nucleotide sequences has become routine. The next chore is to discover what the various sequences do. Some information can be obtained by examining how specific sequences act when placed inside living cells, especially if the sequences encode proteins. But the task becomes massive when one begins to ask detailed questions about how different regions of DNA interact to coordinately control gene expression.

chapter 6

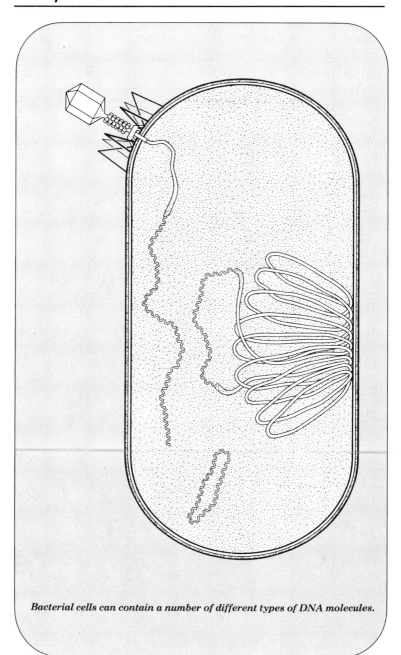

Bacterial cells can contain a number of different types of DNA molecules.

PLASMIDS AND PHAGES

*submicroscopic parasites
as cloning tools*

overview

Plasmids and phages are submicroscopic agents that infect bacteria and use bacterial components to replicate themselves. They contain genetic information, which in most cases is stored in short DNA molecules. Molecular biologists have found plasmid and phage DNA molecules particularly easy to handle and study. Genetic information from animals and other organisms having very long DNA molecules is difficult to study unless the DNA has been cut into small pieces and the pieces have been physically separated. Plasmids and phages assist us in separating DNA pieces. Since DNA fragments can be spliced into plasmid or phage DNA without impairing infectivity, these tiny infectious agents can be used as vehicles to place almost any DNA fragment inside a living bacterial cell. There the DNA fragment reproduces as a part of the plasmid or phage DNA.

INTRODUCTION

Certain types of small DNA molecule are infectious. Once inside a living cell, they can utilize the machinery of the cell to reproduce. According to the biochemist's definition of life, these DNA molecules are not alive; they CANNOT reproduce by themselves. Among other things, they need RNA polymerase from their host (the infected bacterium) to transcribe their DNA into messenger RNAs. They also need host ribosomes to translate the messages into proteins. But even with these deficiencies, infectious DNAs can have profound effects on living cells. Some types take over a cell and kill it, while other types can be beneficial to a cell. These infectious DNA molecules fall into two general types, the **viruses** and the **plasmids**. Viruses surround their DNA molecules with a protective shell of protein; thus, they can sometimes survive for many years outside their host cell. Plasmids, on the other hand, are naked, circular DNA molecules; they are generally found inside cells only.

It is possible to cut DNA molecules from plasmids and **bacteriophages** (viruses that attack bacteria) in a specific place, insert a piece of DNA from another source, and still retain all the information necessary for infection by the plasmid or **phage**. Thus these infectious DNA molecules are useful as tools to transfer DNA from one type of cell to another; they are the cloning vehicles referred to in earlier chapters.

Before delving into what happens when DNA molecules invade cells, it is useful to restate the problem that cloning vehicles help overcome: the specific fragment of DNA one wishes to obtain must be separated from the thousands (or sometimes millions) of other fragments produced during the cutting and splicing process (Figure 5-3), and then the fragment must be located. The separation aspect is not a problem per se; one could place a drop of water containing DNA fragments on an agar plate, smear the drop over the whole surface of the plate, and thus easily separate the fragments. But then it would be very difficult to detect the fragments; they are so small that they are visible only by using an electron microscope, and this instrument is impractical for scanning a large surface or for distinguishing one stretch of DNA from another. Moreover, DNA fragments spread on an agar plate would be too dilute to find with a complementary radioactive probe (described in Chapter 7 and Appendix III). In one sense, such a spreading process would be much like scattering straw from a haystack over a large field. Finding one particular straw would be difficult. The problem with the straw could be solved if only the straw of interest has seeds attached. When the seeds sprouted into a plant, the plant could be seen and the particular straw would be found at the base of the plant.

In the case of the DNA fragments, one needs a way of multiplying the fragments after scattering them; then radioactive probes can be used to find specific ones. Gene cloners multiply the fragments using cloning vehicles and bacterial cells. First, the vehicles and their attached DNA fragments (Figure 5-3) are transferred into bacterial cells (one vehicle and fragment per cell). Next, the cells are scattered on the surface of an agar plate. Third, the cells are allowed to multiply millions of times, forming individual colonies. Each colony contains millions of copies of the cloning vehicle and a particular DNA fragment.

It might appear that cloning vehicles are unnecessary in the process described above. One need only get the DNA fragments into bacterial cells such that each cell obtains but a single fragment; then the cells can be spread out on an agar surface. Each cell will multiply to form a colony, and all the colonies can be tested for the gene of interest. Indeed, many types of bacteria will take DNA in through their cell walls, but very few DNA fragments have the necessary start and stop signals to cause the cellular machinery in bacteria to replicate the DNA fragment. If no DNA replication occurs, the fragment will be diluted out as the bacteria grow and divide, for only the original bacterial cell will contain the DNA fragment. Even after many cell divisions, only one cell in the colony will contain the fragment. To bypass this problem, DNA fragments are first spliced into special pieces of DNA that contain the correct signals for replication. These special pieces of DNA, the plasmid and phage DNAs used as cloning vehicles, then enter the bacterial cells and multiply as the cells multiply.

PLASMIDS

Plasmids are small, circular, double-stranded DNA molecules (Figure 6-1) that occur naturally in bacteria. Like all natural DNA molecules, plasmids contain a special region in their DNA called an **origin of replication**. The origin serves as a start signal for DNA polymerase and ensures that the plasmid DNA molecule will be replicated by the host cell. Many kinds of plasmid have been discovered. Plasmids differ in length and in the genes contained in their DNA. Some of the smaller plasmids, which are popular in gene cloning, have about 5000 nucleotide pairs, enough DNA to code for about five average-sized proteins. In comparison, *E. coli* contains slightly more than four million

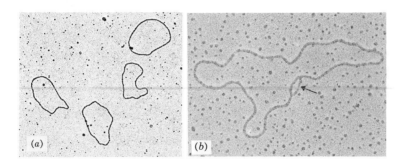

Figure 6-1. Electron Micrographs of Plasmids. *(a) Four DNA molecules of the type called ColEl. These small, circular DNAs are only 0.001 times the length of E. coli DNA (compare with Figure 1-1). Photomicrograph courtesy of Grace Wever, University of Rochester (currently at Eastman Kodak). (b). Enlargement of a plasmid similar to ColEl. Sample preparation conditions were adjusted so that a short region of DNA would become single-stranded (arrow). Photomicrograph courtesy of G. Glikin, G. Gargiulo, L. Rena-Descalzi, and A. Worcel, University of Rochester.*

nucleotide pairs in its DNA, and we have about four billion nucleotide pairs in ours. Many of the larger plasmids are difficult to handle and are transmissible from one bacterial cell to another. Thus they are generally not used in gene cloning; however a brief description of one type is included in Appendix II for interested readers.

An important aspect of plasmid DNA molecules is that they often contain genes that make their host bacterial cell resistant to **antibiotics**. Such resistance turns out to be extremely useful in genetic engineering. For example, in the cloning procedure outlined earlier, DNA fragments are spliced into a plasmid DNA having a gene that confers resistance to **penicillin**. Then the plasmid DNA is added to a culture of bacteria that normally is killed by penicillin. Under the proper experimental conditions, the plasmid DNA enters the cell and multiplies along with the bacterial cell (**transformation**). The bacteria are next spread on an agar plate containing penicillin and incubated for two days. Most of the bacteria are killed. However, the few cells that acquire a plasmid are penicillin resistant, and they grow into colonies. Every colony growing on the agar plate contains cells harboring copies of the plasmid. Thus, when testing colonies for specific genes, gene cloners use antibiotics to avoid examining the millions of bacterial colonies that *fail* to be transformed by a plasmid.

OBTAINING PLASMID DNA

After a gene cloner has spliced DNA fragments into cloning vehicles, transferred the vehicles into bacterial cells, separated the bacterial cells on agar plates, grown the separated cells into colonies, and determined which colony contains the desired DNA, he or she must retrieve the cloned DNA fragment by purifying the plasmid or phage containing it.

The first step in obtaining plasmid DNA is to prepare a liquid bacterial culture containing billions of cells harboring

plasmids (see Chapter 2 and Appendix III for culturing procedure). Each of these bacterial cells contains the cloned DNA
fragment previously identified, and the fragment is still part of
the circular plasmid.

Next the DNA molecules must be removed from living cells.
Bacterial cells that have grown in broth culture are first concentrated; one procedure is to place the broth in a test tube and
allow the cells to gravitate to the bottom of the tube, much like
silt settling out of river water. Generally molecular biologists
speed up the settling process by putting the test tube in a
centrifuge, a machine that works like a merry-go-round. When
a merry-go-round is spinning, a force is generated that throws
objects away from the center of rotation. If a test tube containing
bacteria were attached horizontally on the merry-go-round with
the open end near the center of rotation, the bacteria would be
driven to the bottom of the tube, where they would form a tight
pellet. A centrifuge works in the same way, and the spinning
part that holds the test tube is called a **centrifuge rotor**. The
bacteria stay as a pellet even after the tube is removed from the
centrifuge, so the broth they were growing in is easily poured
off. A small volume of water is then added to the trillion or so
bacteria, and the tube is shaken to distribute the bacteria in the
water. Enzymes and detergents are added to this suspension to
dissolve the cell walls of the bacteria, releasing both bacterial
DNA and plasmid DNA molecules from the cells. At this stage
the content of the test tube is called a **cell lysate**.

Once the DNA molecules have been freed from the cells, the
plasmid DNA must be physically separated from the bacterial
DNA. The two types of DNA differ mainly in their length.
Depending on the particular plasmid, the bacterial DNA may be
up to a thousand times longer. Since it is so large, the bacterial
DNA tends to settle to the bottom of the test tube faster than the
small plasmid DNA. Consequently, one step in the purification
procedure is to put the cell lysate in a tube that is spun in a
centrifuge. The large bacterial DNA will form a pellet in the
bottom of a test tube while the much smaller plasmid DNA will
stay in the upper fluid. Unfortunately, there is usually so much

more bacterial DNA than plasmid DNA that this centrifugation procedure does not completely separate the two kinds of DNA molecules.

The great length of the bacterial DNA, however, makes it possible to carry out an additional type of separation. Bacterial DNA is easily broken by sucking the DNA-containing solution into a **pipette**. This procedure does not break the smaller plasmid DNA; after a few squirts through the pipette, bacterial DNA becomes linear while the plasmid DNA remains circular. Circular DNA has a distinctive property that allows it to be separated from linear DNA. This property is related to the concepts of buoyancy and relative density.

Consider for a moment your own buoyancy in water. If you are totally relaxed and motionless, you tend to float so that only the top of your head is above the surface. You can change your buoyancy in two ways: by putting on a life jacket or by holding some rocks in your hands. The life jacket lowers your overall buoyant density, causing you to float higher, while the rocks increase it, causing you to sink lower. In addition to your own density, the density of the water is important in determining whether you sink or float. For example, water containing a high concentration of salt has a high density, and as a result people easily float in very salty water like the Dead Sea.

In the laboratory, test tubes can be filled with salt solutions so that the salt concentration gradually increases from top to bottom. The gradual change in salt concentration is called a concentration gradient, and it produces a **density gradient**. Salt concentrations can be adjusted so that molecules like DNA will sink until they reach a solution density equal to their own density. At that point the DNA is at **equilibrium**, and if unperturbed, it will remain at that position forever. The important point to remember is that the depth to which a DNA molecule sinks in a density gradient depends on the density of the DNA and the density of the solution.

Now let's return to circular and linear DNA molecules. Both have the same four base pairs, so physically they should have the same density. However, it is possible to add a dye molecule to DNA that acts like a life jacket, lowering the density of the DNA. Linear DNA molecules can bind more dye than circular ones; linear molecules can effectively put on more life jackets and will float higher. Why do more dye molecules bind to linear DNA than to circular DNA? As the dye molecules bind to DNA, they insert themselves between the base pairs and slightly unwind the DNA (Figure 6-2a). Unwinding causes the DNA to twist. Extreme examples of unwinding are shown in Figure 6-2b and c to illustrate the difference between linear and circular DNA. Twists introduced into a linear DNA molecule are quickly lost as the free ends of the molecule rotate over each other (Figure 6-2b). In contrast, twists put into a circular molecule are retained because there are no free ends (Figure 6-2c). As more and more dye molecules bind to circular DNA, the twists accumulate, and it becomes increasingly difficult for subsequent dye molecules to bind. This is not the case with linear DNA. Thus the absence of ends for strand rotation prevents a circular DNA molecule from binding as much dye as a linear one.

Operationally the process is much simpler than the explanation. A DNA preparation is mixed with the dye and a heavy salt in a test tube. At this stage one could rely on the force of gravity to generate the density gradient with the salt water, but it would take a very long time for the salt molecules to accumulate in the

———————————————————————————————————⟶

Figure 6-2. Binding of Dye to Linear and Circular DNA Molecules. (a) *Dyes such as ethidium are flat structures that resemble DNA base pairs. When the dye binds to DNA, it slips in between two adjacent base pairs (shown as blocks). Although the hydrogen bonds holding the base pairs are not broken by the dye, the DNA double helix unwinds slightly (26) degrees per dye molecule bound). (b) Unwinding twists the DNA, but with linear DNA the twisting dissipates as the ends of the DNA rotate. For illustrative purposes, an extreme case of unwinding is shown, and base-pairing has been disrupted. (c) No free ends are present in circular DNA, so twists arising from unwinding accumulate. The twists make further unwinding, and thus dye binding, more difficult. (d) Electron micrograph of twisted plasmid DNA molecules. Photomicrograph courtesy of G. Glikin, G. Gargiulo, L. Rena-Descalzi, and A. Worcel, University of Rochester.*

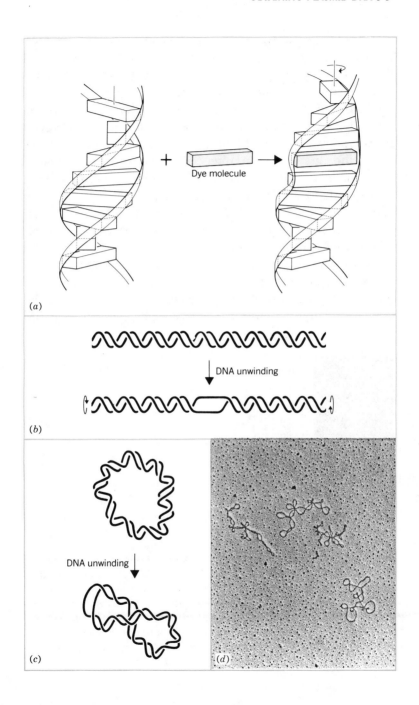

(a)

Dye molecule

(b)

DNA unwinding

(c)

DNA unwinding

(d)

bottom of the tube. It would also take a long time for the DNA molecules to settle to their own density. The time required can be shortened by putting the tube containing DNA, salt, and dye in a centrifuge where forces in excess of 100,000 times gravity can be obtained. After spinning the sample for a day or so, the centrifuge is turned off, and the tube is examined with **ultraviolet light** (black light). When dye molecules bound to DNA absorb the ultraviolet light, they emit a visible, fluorescent light. Two bands can be observed in the tube. The upper one corresponds to the linear DNA (bacterial DNA) and the lower one to the circular plasmid DNA (see Figure 6-3). The circular plasmid DNA can be sucked out of the tube with a pipette, which will then contain pure plasmid DNA.

The cloned genes can be cut out of the plasmid DNA by restriction endonucleases as described in Chapter 5. This produces a small number of DNA fragments; the fragments can be separated from each other by gel electrophoresis (see Figure 5-

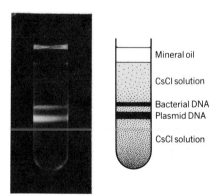

Mineral oil

CsCl solution

Bacterial DNA
Plasmid DNA

CsCl solution

Figure 6-3. Separation of Plasmid and Bacterial DNAs by Dye Bouyant Density Centrifugation. *Plasmid DNA, bacterial DNA, water, dye (ethidium bromide), and a heavy salt (cesium chloride: CsCl) were mixed in a plastic tube and centrifuged for two days at 35,000 rpm. Before centrifugation, mineral oil was added to fill the plastic tube to prevent its collapse from the force of the centrifugal field. After centrifugation, the tube was illuminated with ultraviolet light, and bright orange bands appeared in the tube, indicating the location of the DNA molecules.*

4), the resulting bands can be removed from the rest of the gel by cutting with a razor blade, and the DNA can be washed out of each small gel piece.

BACTERIOPHAGES

Plasmids are naked DNA molecules containing an origin of replication. Bacteriophages, commonly called phages, are more complicated. In addition to having an origin of replication, phage DNA contains genes coding for proteins that form a protective shell around the DNA. But, like plasmids, phages lack the machinery necessary to actually make proteins; consequently, they reproduce only inside living bacterial cells. Both phages and plasmids can be used to separate and amplify specific DNA fragments, but the two have very different means of reproduction. Thus different cloning strategies are employed.

Many phages are like miniature hypodermic syringes (Figure 6-4). The phage DNA is wrapped into a tight ball inside a headlike structure made of proteins. A tail, also made of proteins, is attached to the head. When such a phage particle comes in contact with a bacterial cell, the phage tail sticks to the cell wall, and the DNA is squirted out of the head, through the tail, and into the bacterium (see frontpiece, Chapter 6). Soon after the phage DNA gets into the cell, it begins to take control. Special phage genes are transcribed by the bacterial RNA polymerase, and the resulting messenger RNAs are translated into phage proteins using the bacterial ribosomes. At early stages of infection some phages produce proteins that destroy the bacterial DNA, chopping it into individual nucleotides. Once that has happened, the bacterium is doomed, because all the information needed for its own reproduction is gone. Some phages have genes that produce an RNA polymerase, so they do not have to rely on the host polymerase to make messenger RNA from phage genes.

Many phages also have genes for their own DNA replication

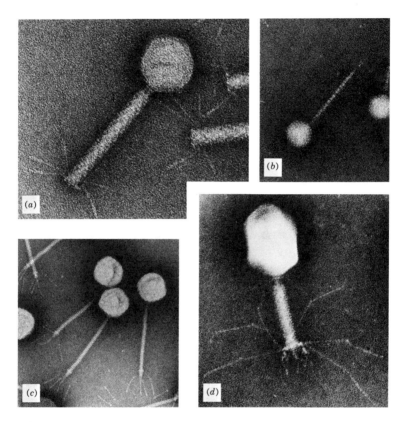

Figure 6-4.   Electron Micrographs of Bacteriophages. *(a) Bacteriophage P2, magnification 226,000 times. (b) Bacteriophage lambda, magnification 109,000 times. (c) Bacteriophage T5, magnification 91,000 times. (d) Bacteriophage T4, magnification 180,000 times. Photomicrographs courtesy of Robley Williams, University of California, Berkeley.*

machinery. When this apparatus is in place, the phages can use nucleotides released from the bacterial DNA to make phage DNA. Hundreds of copies of the phage DNA are made, and within minutes other genes on the phage are turned on to produce new head and tail proteins. The head proteins assemble into heads, phage DNA is packaged inside them, and a tail is attached to each head. The assembly of new phages occurs *spontaneously,* and the total time from injection of DNA to production of new phages can be less than 20 minutes. The bacterium becomes little more than a shell containing hundreds of new phage particles. As a final act, the phage produces an enzyme that destroys the bacterial cell wall, releasing the phage particles to seek new hosts.

Phage efficiency is awesome. One phage produces hundreds of **progeny** particles. Each progeny particle can infect a bacterial cell and produce several hundred more phage particles. By repeating the infection cycle just four times, a single phage particle can lead to the death of more than a billion bacterial cells. Next, consider what happens if a small number of phage particles is added to a *dense* bacterial culture, and before the first bacterial cell is broken open by the newly made viruses, the culture is spread on an agar plate. Within 20 to 30 minutes, the first infected bacteria break open, releasing phage particles. So many bacteria are on the agar plate that the new viruses quickly attach to nearby bacteria and repeat the infection process. Gradually the circle of death expands away from the point where the original infected cell had fallen on the agar surface. Meanwhile, the uninfected cells, which are the vast majority, continue to divide. They are unaffected by the fact that a few of their number are infected, and as they become more numerous, they begin to deplete the food supply. Eventually, the uninfected bacteria completely cover the agar surface except for the small regions where the phages are attacking the cells. Within a day the bacteria stop growing, and their biochemical machinery shuts down. The phages, too, stop reproducing, since they rely on active bacterial machinery to supply energy for their enzymes. Thus the agar plate is covered by a lawn of bacteria containing small holes wherever the phages had been multiplying (Figure 6-5). These holes, called **plaques**, are about an eighth of an inch in diameter, and each one arose from a single phage particle.

If a DNA fragment is spliced into a phage DNA molecule without destroying important phage genes, the phage will reproduce the fragment along with its own DNA when it infects

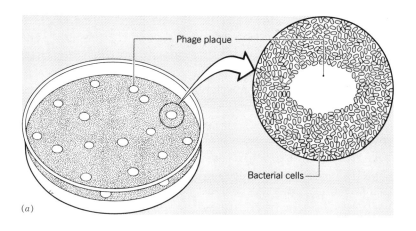

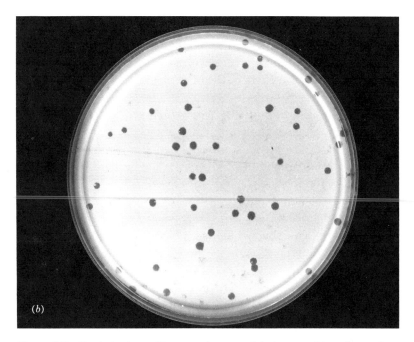

Figure 6-5.   Bacteriophage Plaques. *An agar plate is covered by a lawn of bacteria. The holes in the lawn are regions where phages have killed the bacteria. These holes are called plaques. (a) Schematic diagram. (b) Photograph courtesy of Robert Rothman, University of Rochester.*

a bacterial cell. Gene cloners can determine which plaque has a particular piece of DNA by using a radioactive probe to test the plaques for DNA having base pairs complementary to a specific gene. Once the right plaque has been found, a procedure similar to that described previously for bacterial colonies is used to obtain large amounts of the gene.

First, a piece of sterile wire is poked into the plaque containing the cloned gene. A small number of virus particles will attach to the wire. When the wire is stuck into a fresh culture of bacteria, the phages drop off the wire, attack the bacteria, and multiply billions of times. Large amounts of phage DNA will be made and packaged. Since the cloned gene is actually a part of the phage DNA, it too will be very abundant; moreover, it will be packaged inside a phage head along with the phage DNA sequences. Phage particles can be easily purified by density gradient centrifugation in a way similar to that described earlier for purification of plasmid DNA. Since DNA and protein have different densities, phages, which are a combination of DNA and protein, will have a buoyant density between that of pure DNA and pure protein. Thus, no dyes are needed to separate phages from other cellular components, including bacterial DNA. A tube containing the phage in a heavy salt solution is centrifuged until a density gradient is established and the phage sediments to its own density. When the tube is removed from the centrifuge and examined, the phage will appear as an opalescent band that can be easily sucked out with a pipette (Figure 6-6).

One of the phages used for cloning is called **lambda**. Lambda is slightly more sophisticated than the phages just described. When lambda DNA is injected into a bacterial cell, it has two choices. It can behave as described above and destroy the bacterium, or it can take up residence in the cell. When the latter choice is made, the lambda DNA inserts into the bacterial chromosome; it becomes part of the bacterial DNA (Figure 6-7). The phage genes that normally would produce the proteins to kill the cell are turned off by a repressor protein made from a phage gene. Thus every time the bacterial DNA replicates,

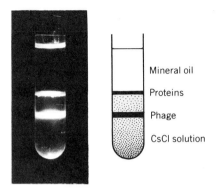

Mineral oil

Proteins

Phage

CsCl solution

Figure 6-6. Purification of a Bacteriophage by Centrifugation. *A bacterial lysate containing phage particles was placed in a plastic tube, mixed with cesium chloride, and centrifuged at 25,000 rpm for one day. During the centrifugation the phage particles formed a band as indicated in the figure. The band above the phage is composed of bacterial proteins and cell wall material. Above this band is a layer of mineral oil used to fill the tube to prevent collapse during centrifugation.*

lambda replicates. In this dormant state lambda DNA does little more than produce repressor protein to keep its genes shut down. At the same time, the repressor protects the bacterial cell from infection by other lambda phages—when the newcomers inject their DNA, their DNA is soon bound by a repressor produced from the resident lambda DNA. Thus the incoming DNA is unable to initiate a **lytic infection** that would kill the cell. Consequently, one can easily find **lysogens** (bacterial cells that are being protected by a resident phage) by looking for bacterial colonies in the middle of a phage plaque.

When a lysogen is made with a phage carrying a cloned gene, a situation arises that is equivalent to one in which a gene is spliced onto a plasmid—the bacterial cell will carry the cloned fragment forever. By constructing the proper regulatory regions on the cloned gene through splicing tricks, it is possible to control when the gene is turned on. Thus either plasmids or phages can be used to insert genes into bacteria. Astronomical numbers of the bacteria can be grown in vats before the cloned genes are induced to produce the desired product.

Cloned genes can be retrieved from lysogens by destroying the phage repressor—the phage DNA then removes itself from the bacterial chromosome and directs the cell to make phage

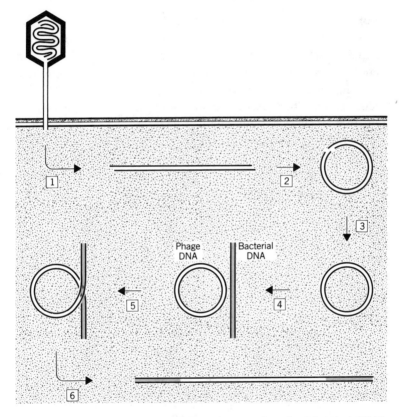

Figure 6-7. Formation of a Lysogen. *(1) Bacteriophage lambda injects DNA through the bacterial cell wall. The resulting linear DNA molecule has sticky ends. (2) The DNA circularizes and (3) becomes ligated. At this point the phage has two choices. It can replicate its DNA, produce progeny phage, and kill the bacterium. Alternatively, it can integrate its DNA into the bacterial DNA (4-6) and remain quiescent for an indefinite number of bacterial generations. Both phage and bacterial proteins play important roles in the integration process.*

particles. These DNA-containing viruses are purified as described earlier (Figure 6-6).

VISUALIZATION OF CLONED DNA

Thus far, little has been said about how a gene cloner really knows that a particular phage or plasmid DNA contains a cloned DNA fragment. Even with positive biochemical and genetic tests, it is reassuring to have physical evidence and, if possible, to actually see the cloned region of DNA. The two methods described below provide an unambiguous demonstration that one DNA fragment has been inserted into another. Figure 6-8 illustrates one way in which a human DNA fragment would change the restriction map of the plasmid into which it had been cloned. In this example, the plasmid contains a single site where a particular restriction endonuclease cuts. At this site human DNA fragments, generated by the same endonuclease, are spliced into the plasmid molecules to form recombinant DNA molecules (step 1, Figure 6-8c). The recombinant DNA molecules are introduced into bacterial cells, individual clones

→

Figure 6-8. Analysis of Recombinant DNA by Gel Electrophoresis. *(1) Plasmid and human DNA molecules are cut and spliced together to form recombinant plasmid DNA molecules. Many combinations of DNA fragments join, producing many different recombinant DNAs (2) The recombinant DNAs are introduced into bacterial cells, and individual colonies are grown. A bacterial colony containing cloned DNA is identified (Appendix III), and recombinant plasmid DNA is isolated from it. (All of the recombinant DNA molecules from a colony are identical. They all have two restriction endonuclease cleavage sites, one at each junction between human and plasmid DNA.) (3) Cleavage of the recombinant plasmid DNA with restriction endonuclease produces two discrete DNA fragments. (4) The original plasmid contains only one restriction endonuclease cleavage site, so it is cleaved only once by the nuclease. (5) The products of steps 3 and 4 are analyzed by gel electrophoresis. The original plasmid produces only one band; the recombinant plasmid produces two.*

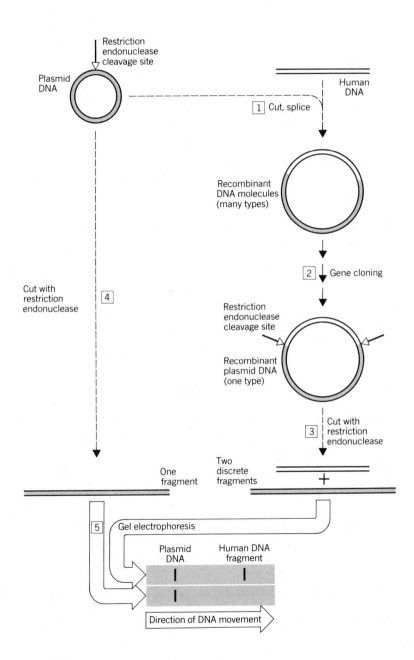

are obtained (Appendix III), and recombinant DNA molecules are purified. This produces a single *type* of recombinant plasmid (step 2, Figure 6-8). When these DNAs are cut with the restriction endonuclease (step 3, Figure 6-8), two pieces of DNA are created. These two pieces can be analyzed and compared with the single piece generated by cleaving the original plasmid (step 4, Figure 6-8) using gel electrophoresis (step 5, Figure 6-8).

The second technique, called **heteroduplex mapping**, combines nucleic acid hybridization (Figure 4-10) with electron microscopy. A hybrid DNA molecule, in which the two strands share regions of nucleotide sequence complementarity but also have regions that are not complementary, will have a distinctive pattern of double-stranded and single-stranded regions. These regions can be distinguished with an electron microscope. The principle of heteroduplex formation is illustrated in Figure 6-9.

Figure 6-10 illustrates a more complicated example. In this case, two DNA fragments have been inserted into separate bacteriophage lambda DNA molecules. The two cloned fragments have regions that are **homologous** (regions of identical nucleotide sequences, a, b, c, d, e, f in Figure 6-10*a*), but fragment i contains an additional set of sequences (x, y, z in Figure 6-10) not present in fragment ii. Whenever a DNA fragment is spliced into a plasmid or phage DNA molecule, the fragment can insert in either of two orientations (Figure 6-10*b*). In our example, fragments i and ii have been inserted into lambda DNA in opposite orientations, and the recombinant DNAs would be diagrammed as shown in Figure 6-10*c*. Note

---

Figure 6-9. Formation of Partially Homologous Hybrid DNA Molecules.
Regions of molecules I and II are identical (a b c d e f); the black area of DNA molecule I (x y z) contains nucleotide sequences not found in molecule II. The two DNAs are heated to separate them into single strands. They are then mixed and cooled to allow the strands to hybridize. Some double-stranded molecules will form that are identical to the original DNA molecules (homo-duplexes). In addition, other combinations of double-stranded molecules will form, among which are the hybrids I-II' and II-I'. These are called heterodu-plexes; the regions x y z and x' y' z' remain single-stranded because their complement is absent in strands II and II'.

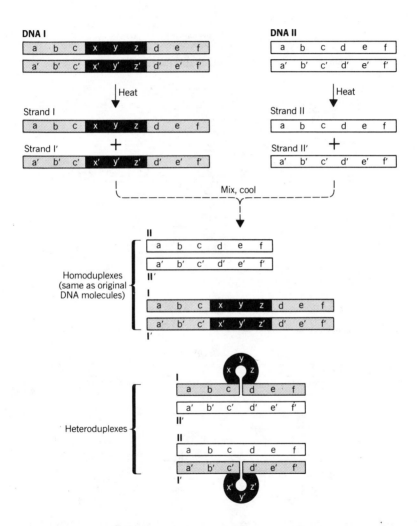

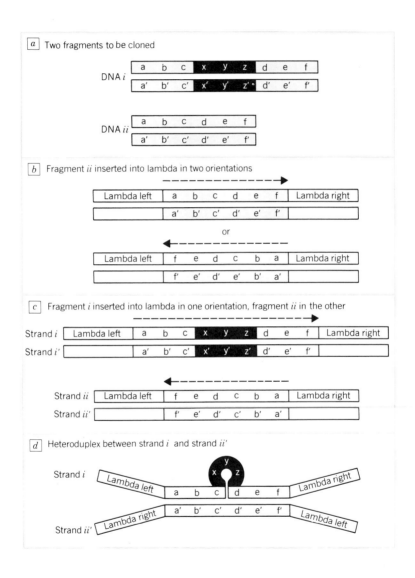

a Two fragments to be cloned

b Fragment *ii* inserted into lambda in two orientations

c Fragment *i* inserted into lambda in one orientation, fragment *ii* in the other

d Heteroduplex between strand *i* and strand *ii'*

that lambda DNA has distinct left and right arms that do *not* share nucleotide sequence homologies. When the two recombinant DNA molecules in Figure 6-10c are hybridized by the procedures illustrated in Figure 6-9, some of the hybrids would look like Figure 6-10d. The complementary regions a b c d e f and a' b' c' d' e' f' would become double stranded, x y z would remain single stranded because strand ii' lacks the x' y' z' region, and the lambda DNA would remain single stranded because the left and right arms are not homologous. Figure 6-11a shows an electron micrograph of such a structure.

PERSPECTIVE

Plasmids and phages are among the smallest and most efficient infectious agents in nature. Some are so efficient that they use the same nucleotide sequence to encode two different proteins. Plasmids have obtained considerable study because of their medical importance: some plasmids carry genes that make their host bacteria resistant to antibiotics. Through our massive use of

---

Figure 6-10. Formation of Heteroduplex Between Two Lambda DNA Molecules Containing Partially Homologous Cloned Sequences in Opposite Orientation. *(a) The diagrammatic representation of two fragments to be cloned into lambda DNA. Fragments i and ii are identical (a b c d e f) except that i contains an extra piece of DNA (x y z). (b) A fragment can insert into lambda DNA in two orientations, thus producing two different types of clone having the same pieces of DNA. (c) Diagrammatic representation of two partially homologous recombinant DNAs (i and ii) in opposite orientation. (d) Representation of one of the possible hybrids that can form between recombinant molecules diagrammed in c. Strand ii' has been turned around so that the homologous regions (a' b' c' d' e' f') will pair with their complement (a b c d e f) in fragment i. When this happens, the lambda DNA will not be able to form base pairs because the left and right arms are not homologous. Thus the lambda DNA remains single-stranded. Molecule i will have an additional short single-stranded region (x y z) because molecule ii lacks its complement (x' y' z').*

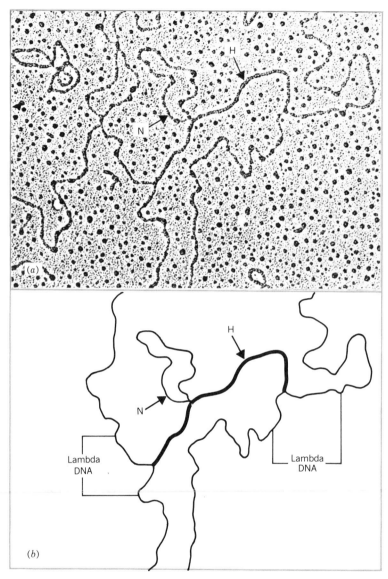

Figure 6-11. A Heteroduplex Formed Between Two Lambda DNA Molecules Containing Partially Homologous Cloned Sequences in Opposite Orientation. (a) electron micrograph. Photograph courtesy of Grace Wever, University of Rochester (currently at Eastman Kodak). (b) An interpretation of the electron micrograph. Heavy line indicated by arrow labeled H represents double-stranded DNA (a b c d e f and a' b' c' d' e' f' from Figure 6-10d. Thin line indicated by arrow labeled N represents single-stranded, nonhomologous region (x y z, Figure 6-10d).

114

these drugs, we have encouraged the spread of plasmids to the point where almost every type of bacterium pathogenic to man now carries these infectious drug-resistance factors. We can no longer rely on antibiotics to cure our diseases.

Phages are important in another way. For three decades molecular biologists concentrated on discovering how phages regulate their genes and replicate their DNA. As a result, phage studies provide the basis for most of our understanding of DNA. As our studies shift to complex organisms, plasmids and phages are assuming a new role in biology, that of the work-horse harnessed to move genes from one organism to another.

chapter 7

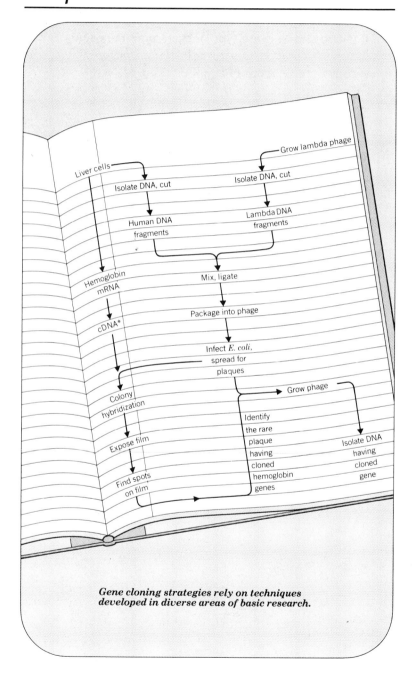

Gene cloning strategies rely on techniques developed in diverse areas of basic research.

CLONING A GENE

*isolation of a
hemoglobin gene*

overview

*Gene cloning procedures involve splicing human DNA
fragments into phage or plasmid DNA molecules. When
phages are used, the recombinant DNA molecules are
packaged inside phage particles so that they can infect bacteria and generate phage plaques. Thousands of phage
plaques are then tested for the presence of a specific gene
by nucleic acid hybridization using a highly radioactive
probe.*

*Gene cloning strategies with plasmids often use the
drug-resistance properties of plasmids to help locate bacterial colonies into which human genes have been cloned. In
one type of strategy, plasmids having genes for resistance
to two different drugs are chosen. One drug-resistance
gene contains a restriction endonuclease cleavage site
where DNA fragments can be inserted; insertion of a fragment into this site on the plasmid destroys the drug-
resistance gene. The second drug-resistance gene is not af-*

fected. Thus recombinant plasmid DNA molecules will confer resistance to the second drug only. When bacterial cells are transformed with plasmids thought to be recombinants, colonies are obtained that grow on agar containing the second drug. Each of these colonies is then tested for growth on agar containing the first drug; those that fail to grow are saved and are tested for the presence of a specific gene by nucleic acid hybridization using a highly radioactive complementary DNA probe.

INTRODUCTION

Gene cloning is much like baking a cake in that recipes are available for both processes. This chapter describes a recipe for cloning hemoglobin genes to summarize and tie together the concepts developed in the preceding chapters. A few technical details have been added to complete your understanding of gene cloning. The original hemoglobin recipe is particularly instructive because it uses both plasmids and phages for cloning.

Some genes are medically important, while others serve primarily as subjects for biological research. The genes coding for **hemoglobin** fall into both categories. Hemoglobin is the blood protein responsible for moving oxygen from our lungs to our body tissues. One of the medically important disorders caused by improper hemoglobin function is the serious disease called sickle-cell anemia. From a research point of view hemoglobin genes are interesting because a number of genes code for different hemoglobins that function at different stages of our lives. Biologists want to understand how one hemoglobin gene is switched on and how another is switched off. Having the hemoglobin genes cloned makes many new experiments possible.

OBTAINING DNA

The first step in the recipe is to find a source for hemoglobin genes. The initial studies were done with rabbit genes, but similar procedures would work with human genes. With few exceptions all the cells in a rabbit body contain identical DNA molecules. Thus the DNA from virtually any rabbit cell contains hemoglobin genes, and any type of body tissue can be ground up to obtain these genes. Here the threshold of life is crossed. As seen with phages, DNA can contain information without being part of a living organism. Thus, hemoglobin genes can be obtained from a live or dead donor, and they can be stored for an indefinite time in a frozen state.

Now consider the recipe used in the pioneering work on rabbit genes. First, DNA was purified from rabbit livers. Liver tissue was frozen, placed in a blender, and ground until the cells were broken. The molecules held inside the cells were released, and a detergent solution was added, along with an enzyme that breaks down proteins. (These two agents remove proteins that may be sticking to the DNA and also inactivate nucleases present in the extract that might otherwise begin chopping the DNA.) The cell lysate was incubated for several hours, whereupon it was treated with **phenol** to remove proteins that had escaped enzymatic digestion. Finally, the DNA-containing solution was centrifuged in a density gradient as described in Chapter 6 to purify the DNA. Only one band formed in the centrifuge tube, and it was easily sucked out with a syringe and placed in a test tube. Hemoglobin genes, millions of them, were in the test tube. But so were millions of copies of a hundred thousand other genes.

CLONING INTO BACTERIOPHAGE LAMBDA DNA

Hemoglobin genes were initially separated from all the other genes in the DNA preparation by cloning them into bacteriophage lambda DNA. In this part of the recipe lambda phage particles were purified by a method similar to that described in Chapter 6 (Figure 6-6). Next, DNA was extracted from the phage particles. Since it is necessary to shorten the lambda DNA to make room for the rabbit DNA inside the virus particle, the lambda DNA was cut into several pieces by a restriction endonuclease. The two DNA fragments required by lambda to infect a bacterial cell were spliced together, and nonessential fragments were discarded. The rabbit DNA was broken into large pieces, and the two types of DNA were mixed. Ligase was added to splice the fragments, which were then coated with phage proteins. Thus the recombinant DNAs, in many different combinations, were neatly packaged inside phage particles. In this form they were easily transferred into bacterial cells by the mechanism normally used by the phage to inject its DNA. Often, the packaged phage DNA contained a piece of rabbit DNA.

The viruses formed *in vitro* (i.e., in a test tube) were allowed to infect *E. coli* cells, which subsequently were spread on agar. This spreading step separated the phage particles from each other. Phage plaques formed in the lawn of bacteria as each phage multiplied (Figure 6-5). Each plaque arose from a different phage particle, and each one was composed of millions of identical phage particles. Unfortunately, very few contained a hemoglobin gene. The next part of the recipe was designed to find the rare plaque that did.

OBTAINING A RADIOACTIVE PROBE

The general procedure was to screen all the plaques to determine which one had hemoglobin DNA. This was done with a radioactive nucleic acid probe (see Figure 4-10), relying on the principle of complementary base-pairing as discussed in earlier chapters. The first problem was to obtain a nucleic acid, RNA or DNA, containing the base sequences of a hemoglobin gene. Since cells make hemoglobin messenger RNA faithfully from the genes, a logical first choice was to isolate and purify hemoglobin messenger RNA. In most cases of gene cloning, however, it is very difficult to isolate messenger RNA that represents the information from just a single gene. Generally, cells make many types of messenger at the same time, and the messenger RNA molecules are chemically and physically too similar to be separated. But hemoglobin represents a special case, for red blood cells produce mainly hemoglobin. Most of the messenger RNA molecules in these cells are hemoglobin messengers; consequently, red blood cells proved to be a good source for hemoglobin messenger RNA.

Blood was collected, and the red blood cells were concentrated by centrifuging them so that they formed a pellet in the bottom of a test tube. The pellet of red blood cells was resuspended in a small volume of water and salts, and the cells were broken by adding detergents. The lysate was treated with phenol to help remove proteins, and alcohol was added. Alcohol caused RNA to form a white precipitate that was separated from other cell components by centrifugation. At this stage the RNA was pure, but the sample included ribosomal and transfer RNA as well as hemoglobin messenger RNA.

Hemoglobin messenger RNA was purified by taking advantage of a feature unique to many kinds of messenger RNA found in higher organisms: these RNAs often have several hundred A's attached to one end. A glass column, similar to that illustrated in Figure 4-9, was filled with single-stranded DNA composed only of T's attached firmly to cellulose. The RNA mixture was passed through the column. As the RNA percolated through, the stretches of A's on the hemoglobin messengers formed complementary base pairs with the long runs of T's fixed to the cellulose. The complementary base pairs were strong enough to prevent the hemoglobin messengers from flowing through the column. The other RNA molecules did flow through, and they were thrown away. The hemoglobin messengers were then removed from the column by breaking the base pairs, a process that occurs after gentle warming.

At this stage the messenger RNA was almost ready to use to

identify the phage plaque containing the hemoglobin gene. But first the information in it, the sequence of nucleotides, had to be converted to a highly radioactive form. This could have been done by mixing together RNA, radioactive nucleotides, and the enzyme called **reverse transcriptase**, an enzyme obtained from tumor virus particles by combinations of manipulations similar to those outlined in Chapter 4. Reverse transcriptase makes DNA from free nucleotides, using the RNA as a template. The DNA, called complementary DNA (cDNA), can be made highly radioactive if its components, the nucleotides, are radioactive.

In many examples of gene cloning, radioactive complementary DNA is suitable to locate plaques containing cloned genes. However, in the initial cloning studies with hemoglobin it was necessary to screen hundreds of thousands of phage plaques. Thus, large amounts of complementary DNA were needed. It was decided to first synthesize complementary DNA, convert the cDNA into double-stranded DNA using DNA polymerase, and then clone it into a plasmid to obtain large amounts. The cloned DNA would then be made radioactive. So, the double-stranded DNA copies of hemoglobin mRNA were spliced into plasmids, and the resulting recombinant DNA molecules were transferred into *E. coli* cells. The cells were grown into colonies, and these colonies were tested for the presence of hemoglobin nucleotide sequences using small amounts of radioactive hemoglobin messenger RNA. Some of the details of this procedure are presented below. A more general procedure for cloning with plasmids is described in Appendix III.

To insert DNA copies of hemoglobin mRNA into plasmid DNA, plasmid DNA was cut once with a restriction endonuclease to convert the circular DNA into a linear molecule (Figure 7-1*a*). Long single-stranded tails were produced by treating each DNA with an enzyme called lambda exonuclease (Figure 7-1*b*). Another enzyme was then used to put about 100 T's onto the tails of the DNA copies and about 100 A's onto the ends of the plasmid DNA (Figure 7-1*c*). When the two DNAs were mixed, the A's and T's formed base pairs, joining the DNA copies to the

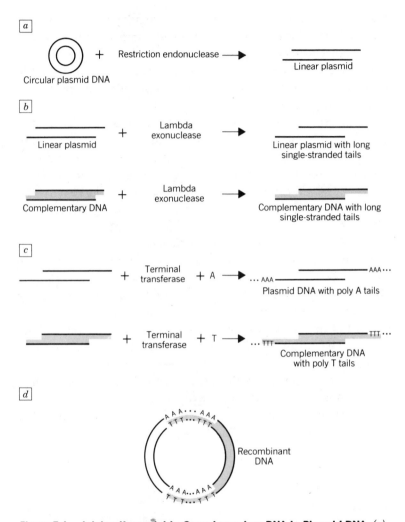

Figure 7-1. Joining Hemoglobin Complementary DNA to Plasmid DNA. *(a) Plasmid DNA is converted from a circular molecule to a linear one. (b) Single-stranded tails are created enzymatically on plasmid DNA and on a hemoglobin DNA copy by treating the DNA molecules with an enzyme called lambda exonuclease. (c) Another enzyme, terminal transferase, adds complementary nucleotides to the tails. When the plasmid and hemoglobin DNA copy are mixed, the complementary tails form base pairs producing a circle (d).*

plasmids (Figure 7-1*d*). The resulting circular DNA molecules were added to a culture of *E. coli* cells, and the recombinant DNA molecules (cDNA plasmids) entered some of the cells. The plasmid DNA contained a gene conferring resistance to the antibiotic tetracycline, so when the *E. coli* cells were placed on agar plates containing the drug, only cells containing the plasmid grew into colonies. The colonies were tested for the presence of hemoglobin DNA by the nucleic acid hybridization method described at the end of Chapter 4 (also see Appendix III, Figure AIII-4). Eighty percent of the colonies contained hemoglobin DNA, and they became the source for the large amounts of DNA needed to search for phages containing hemoglobin DNA.

The cDNA plasmids were radioactively labeled using an enzyme that could add a radioactive phosphorus atom to the end of DNA, and they were used to test 750,000 phage plaques to discover plaques containing hemoglobin genes. The testing procedure involved the nucleic acid hybridization method described at the end of Chapter 4 and was similar to the strategy used to test bacterial colonies (Figure AIII-4). Four independent phage plaques were found to have hemoglobin genes. The nucleotide sequences in these cloned hemoglobin genes were determined by methods similar to those described in the next chapter. This type of study has led to surprising findings, some of which are outlined in Chapter 9.

PERSPECTIVE

Often existing technology dictates the direction taken by biological research. For example, in the mid-1970s methods were available for isolating hemoglobin messenger RNA from red blood cells, cells specialized to make hemoglobin. These messenger RNA molecules could be used as templates for the synthesis of radioactive probes to locate cloned hemoglobin genes. Other genes, which might have been just as interesting to study as hemoglobin, are usually transcribed into RNA in cells that make messengers of many different types. Although messenger RNAs can be purified as a group, they are so similar that they are not easily separated from each other to make probes for particular genes. Thus hemoglobin genes were an obvious choice for these early cloning studies.

The recipe for cloning hemoglobin genes also illustrates how biological research builds on previous developments. Messenger RNA isolation, phage and plasmid manipulation, restriction endonuclease cutting of DNA, enzymatic synthesis of DNA, and nucleic acid hybridization were all highly refined technologies being used for other studies. Gene cloners combined them to serve a new purpose. The same principle is seen in the next chapter, which discusses strategies for using cloned genes.

chapter 8

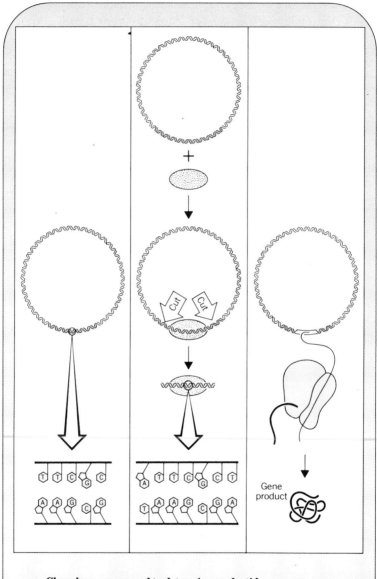

Cloned genes are used to determine nucleotide sequences, to identify sites where proteins bind to DNA, and to produce large amounts of gene products.

chapter 8

USING CLONED GENES

determining nucleotide sequences, walking along DNA, and binding proteins to DNA

overview

Gene cloning technologies make it possible to obtain the large amounts of specific regions of DNA necessary to study gene structure and function. Gene structure is analyzed by determining the nucleotide sequence of a section of DNA. A short region of DNA is first purified and cloned into a bacterial virus called M13. DNA polymerase is used to produce a series of DNA fragments that extend from a specific spot in M13 DNA to particular nucleotides in the sequence being analyzed. By measuring the lengths of these fragments, it is possible to determine the position of each nucleotide relative to the fixed spot in M13 DNA. The nucleotide sequence can then be deduced. The exact position of a gene can be located by finding a nucleotide sequence identical to that predicted from the amino acid sequence of the protein product of the gene.

127

Gene function is studied by examining how the protein product of a gene interacts with other proteins, with DNA, and with small molecules. Gene cloning makes it possible to produce the large amounts of particular proteins and particular regions of DNA needed for this type of study.

INTRODUCTION

In the first seven chapters we focused on the concepts of molecular biology to explain the process of gene cloning. It may have appeared that cloning a particular gene is an end unto itself. It is not. Gene cloning is but a tool, albeit a very powerful tool, that biologists use to explore life. In this chapter we examine several uses of gene cloning. Then in the final chapter we consider some of the surprising findings that have recently emerged.

Gene cloning is a way to obtain large amounts of short, specific regions of DNA. These sections of DNA that are dissected out can be very short, often representing less than one millionth (0.000001) of the total nucleotide sequence of an organism's DNA. By using gene cloning we can obtain enough molecules of DNA and proteins to study gene structure, function, and regulation at the level of molecular interactions.

ANALYSIS OF GENE STRUCTURE

The initial step in analyzing gene structure is to determine the nucleotide sequence of the gene. DNA molecules are incredibly long and monotonous; deciphering the exact order of thousands of A's, T's, G's and C's requires an ingenious combination of many of the molecular tools discussed in earlier chapters.

The recipe outlined in Chapter 7 makes it possible to insert a particular DNA fragment—for example, a fragment of human DNA—into either plasmid DNA or phage DNA. The resulting recombinant DNA is copied trillions of times by bacteria, and these molecules are purified. Further analysis requires that the human DNA be separated from the cloning vehicle DNA (the plasmid or phage DNA). Restriction endonucleases and gel electrophoresis are used in the following way to accomplish this. Often the human DNA fragment is inserted into the cloning vehicle at a site where a specific restriction endonuclease cuts; if the splicing event regenerates the recognition site for the enzyme (Figure 8-1), the recombinant DNA need only be treated with the same restriction endonuclease to liberate the human DNA fragment from the cloning vehicle (see Figure 6-8). Since the cloning vehicle and the human DNA fragment usually have different lengths, they can be physically separated into discrete

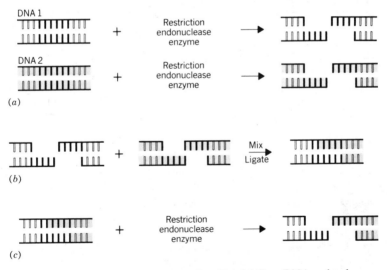

Figure 8-1. Regeneration of a Restriction Site. *(a) Two DNA molecules containing a recognition site for the same restriction endonuclease (solid regions) are cut with that enzyme. If the DNA molecules are ligated together (b), the restriction site is regenerated. The ligated DNA can in turn be cut into fragments (c) by subsequent treatment with the restriction endonuclease.*

bands by gel electrophoresis. Bands of DNA can be removed from the gels easily. Each band contains a large number of small, *identical* DNA fragments. If necessary, these fragments can be cut into still smaller fragments with other restriction endonucleases, and these small fragments can also be separated by gel electrophoresis (Figure 8-2).

The general strategy for determining the nucleotide sequence of a DNA fragment involves establishing a reference point and then measuring the distance from the reference point

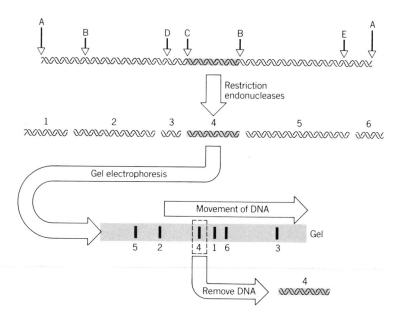

Figure 8-2. Cutting DNA into Pieces of Manageable Size. *A cloned human DNA fragment is often much longer than the region being sought (shaded region); thus, it must be cut into smaller pieces. Arrows A represent the ends of the human DNA fragment, and arrows B, C, D, and E indicate cleavage sites for four different restriction endonucleases. Treatment of the human DNA fragment with these four nucleases produces fragments 1 through 6; these fragments are physically separated by gel electrophoresis. Each band in the gel (Figure 5-4) results from billions of identical DNA molecules. Individual bands are cut out of the gel, and the DNA is removed. In the example shown, band 4 would be saved for further study.*

to each of the nucleotides of a known type, for example, to each adenosine (A). In one method, billions of copies of a single type of DNA fragment are treated with a large number of DNA polymerase molecules, and new DNA is made under conditions in which all new molecules start at one position in the nucleotide sequence but they terminate at many different positions, always at an A (step 1, Figure 8-3). By making a large number of new DNA molecules, one can generate a collection of new DNA fragments that contain all the possible lengths from the reference point to the A's. Measurement of these lengths by gel electrophoresis (step 2, Figure 8-3) establishes the exact distance

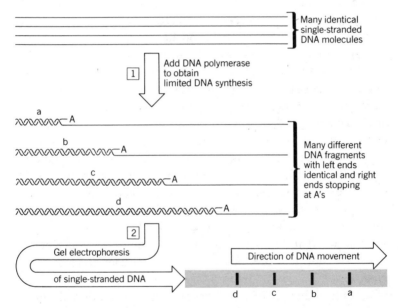

Figure 8-3. Creating and Analyzing a Collection of DNA Fragments Having Different Lengths. *(1) Many identical single-stranded DNA molecules are partially replicated. A collection of shorter molecules (a–d) is formed. All the new DNAs have the same left end but different right ends, which in this example always stop at an A. (2) The collection of fragments is denatured (converted into single strands) and analyzed by gel electrophoresis to determine the length from the left end to the position of each A (the distance a DNA molecule moves in the gel during electrophoresis is related to its length).*

of each A in the nucleotide sequence from the reference point. After this procedure has been repeated for each of the other three nucleotides (T, C, and G), it is possible to deduce the nucleotide sequence.

A few more details must be added to explain how DNA polymerase is started and stopped at the correct places. First, the fragment to be sequenced is spliced into a particular site in the DNA of a phage called M13 using standard cutting and splicing methods (Figure 8-4). The recombinant phage DNA is used to infect bacterial cells, and virus particles are produced. These particles contain only one of the two strands of DNA (M13 is a single-stranded, DNA-containing virus). The single strands are purified from the viruses. They are then mixed with short, single-stranded pieces of DNA that are complementary to a region of M13 DNA near the position where the gene to be sequenced has been inserted. The short piece forms base pairs with the M13 DNA, creating a short double-stranded region of DNA that can serve as a **primer** for synthesis (Figure 8-5*a*). A large number of these partially double-stranded DNA molecules are mixed with DNA polymerase and radioactive nucleotides. DNA polymerase synthesizes radioactive DNA from the M13 template, beginning at one end of the short, double-stranded region (DNA polymerase requires a primer to start synthesis of DNA; thus, it starts synthesis only where the primer is located). DNA polymerase soon crosses into the human DNA region and uses it as a template to make new DNA. The polymerase can be stopped by including a nucleotide analogue (a fake nucleotide) in the reaction mixture, one that cannot be used to extend the growing DNA chain. If the analogue behaves like an A, the new, radioactive chain will stop after it would normally have added an A (Figure 8-5*b*). Normal A's are included in the reaction mixture to allow some DNA synthesis before an A analogue stops the reaction. Since there are many A's in a nucleotide sequence, the stops will not always be in the same place. Thus, by carefully adjusting the amount of analogue in the reaction mixture, it is possible to create a collection of radioactive DNAs that begin at a distinct spot (the end of the

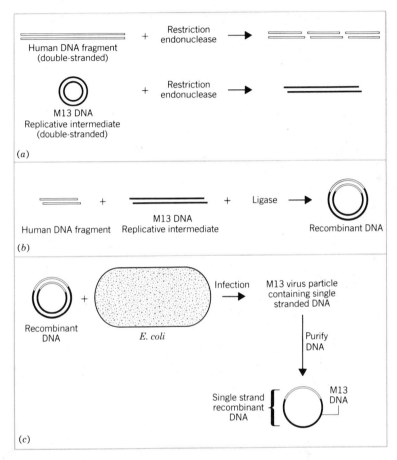

Figure 8-4. Cloning into Phage M13. *(a) A restriction endonuclease is used to cut a human DNA fragment and a purified double-stranded replicative intermediate of phage M13 (single-stranded DNA viruses form double-stranded DNA molecules as a part of their life cycle). (b) The DNAs are mixed and ligated to form a recombinant DNA. (c) The recombinant DNA is used to infect* E. coli *cells, producing M13 virus. The virus particles contain only one of the two DNA strands. This DNA strand, which is the same for all the virus particles, can then be purified in large quantities.*

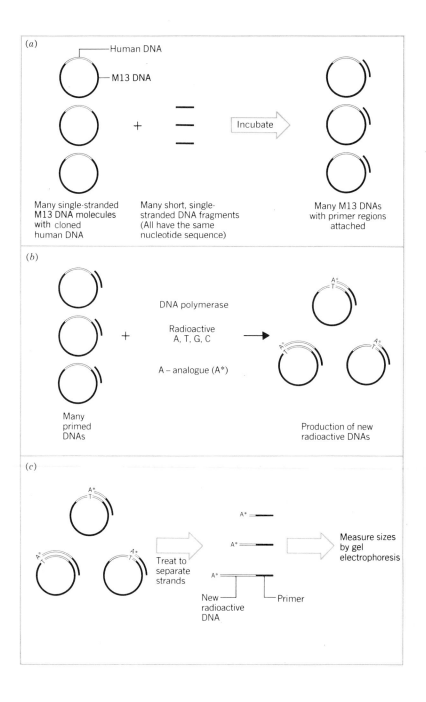

primer) and stop at the various positions where A's occur. The identical procedure is repeated in three other test tubes using analogues for G, C, or T. Thus four separate collections of molecules are made. Every molecule starts at the same place, and the various types end at different distances from the starting point.

The lengths of the molecules are measured by gel electrophoresis (Figure 8-5c) under conditions that are so precise that they can separate DNA molecules (created using A, T, C, or G-analogues) differing by only one nucleotide. The four collections of molecules are electrophoresed next to each other, and after electrophoresis a piece of film is exposed by the radioactivity in the DNA. By examining only radioactive molecules, it is not necessary to consider the many nonradioactive DNA fragments that might otherwise complicate the experiments. A series of bands is seen (Figure 8-6). The lowest band in Figure 8-6 represents molecules that extend to an A, for it appears in the sample containing the A analogue (A*). The next higher band is in the G-analogue lane, so the next nucleotide is a G. Thus nucleotide sequences can be read directly from pictures like that shown in Figure 8-6.

The nucleotide sequences from a number of adjacent restriction fragments can be fit together to produce very long sequences. Recently the entire sequence for bacteriophage lambda DNA (48,498 nucleotides) was determined in this way. Particular genes are located in the sequence by comparing the nucleo-

---

Figure 8-5. Creating a Collection of Variable Length Copies of Cloned DNA. *(a) Many single-stranded M13 recombinant DNA molecules from Figure 8-4 are mixed with short single-stranded DNA fragments complementary to a region of M13 DNA near the junction between M13 DNA and the cloned DNA. The fragment forms a double-stranded region with M13 DNA. (b) The short double-stranded region acts as a primer for DNA polymerase. New DNA is synthesized from radioactive A's, T's, G's, and C's onto one end of the primer. An analogue of A (A*) is added to halt synthesis at the various positions where its complement, T, occurs in the nucleotide sequence of the cloned DNA. (c) The DNAs are treated so that all of them become single-stranded. Their sizes are then measured by gel electrophoresis.*

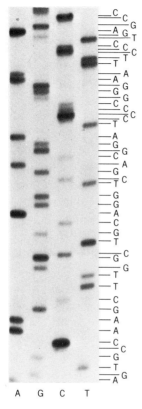

Figure 8-6. DNA Sequencing Gel. *Four radioactive DNA samples were electrophoresed in adjacent lanes. The four samples labeled A, T, C, and G across the bottom of the figure were synthesized in the presence of analogues to A, T, C, and G, respectively, as described in Figure 8-5. The nucleotide sequence of the DNA is indicated on the right. Photograph courtesy of Richard Archer, Janice Zengel, and Lasse Lindahl, University of Rochester.*

tide sequence to that expected from the amino acid sequence of the protein made from the gene. For example, if the left-most amino acid in the protein is methionine, the second one from the left is tryptophan, and the third one is phenylalanine, we expect the nucleotide sequence of that portion of the gene to be A-T-G-T-G-G-T-T (T or C) because we know the triplet codon for each amino acid. The ninth nucleotide could be either T or C because phenylalanine is encoded by two triplets, T-T-T and

T-T-C. This type of comparison allows us to determine the exact position of a gene. Biologists are now carefully examining nucleotide sequences both inside and outside genes to try to understand how genes are turned on and off.

WALKING ALONG DNA

Once a piece of DNA has been cloned, determining the sequence is straightforward. Often molecular biologists want to know the nucleotide sequence of adjacent regions of DNA. This information is obtained by a modification of the cloning procedure called **walking along DNA**. The general strategy is to use a nucleotide sequence near one end of the cloned region as a probe for locating adjacent, overlapping regions in a collection of recombinant DNAs (Figure 8-7). Two restriction endonucleases are used; first, one is found that cuts at or near an end of the cloned fragment (X in Figure 8-7a) and another is found that cuts at a site in the cloned sequence and also at a site far outside the cloned region (R in Figure 8-7a). When human DNA is cut with enzyme R only and the fragments are cloned into plasmids, one type of recombinant DNA (region II, Figure 8-7a) will partially overlap with the original cloned sequence. The overlapping sequence is indicated as a solid region in Figure 8-7. Bacterial colonies containing this DNA are identified by the colony hybridization technique (Figure AIII-4) using a radioactive probe from the overlapping region obtained as outlined in Figure 8-7a. The DNA fragment (region II, Figure 8-7) can be isolated from recombinant plasmids by methods described previously and the nucleotide sequence of this region can be determined. By repeating this process, one can move along a DNA molecule, successively cloning and sequencing small bits of DNA.

ANALYSIS OF GENE FUNCTION

The protein products of many important genes are present in very tiny amounts inside living cells, and it is often very difficult to obtain enough of a particular protein to study its properties

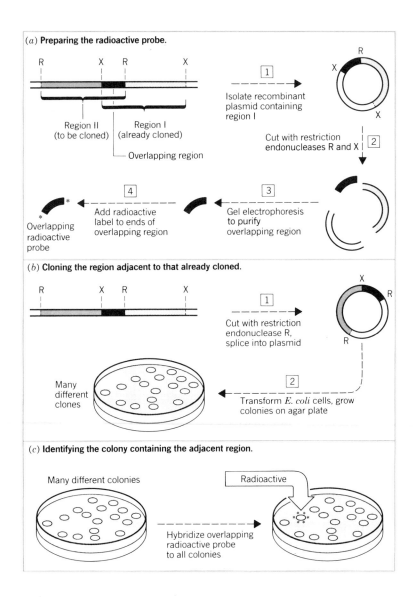

(*a*) **Preparing the radioactive probe.**

Region II (to be cloned)   Region I (already cloned)

Overlapping region

1 Isolate recombinant plasmid containing region I

Cut with restriction endonucleases R and X 2

3 Gel electrophoresis to purify overlapping region

4 Add radioactive label to ends of overlapping region

Overlapping radioactive probe

(*b*) **Cloning the region adjacent to that already cloned.**

1 Cut with restriction endonuclease R, splice into plasmid

2 Transform *E. coli* cells, grow colonies on agar plate

Many different clones

(*c*) **Identifying the colony containing the adjacent region.**

Many different colonies

Radioactive

Hybridize overlapping radioactive probe to all colonies

and interactions with other molecules. But in a number of cases gene cloning has erased this problem. Genes for particular proteins have been purified by cloning technologies, and these genes have been used to direct cells to produce large amounts of the proteins.

Some of these protein products are particularly interesting because they act on DNA. Repressors are an example. Chapter 3 pointed out that repressors prevent RNA polymerase from transcribing RNA from certain genes, and thus repressors regulate gene expression. Understanding this aspect of gene control involves knowing exactly where the repressor binds to the DNA. Gene cloning technologies have made it possible to obtain large amounts of both repressor protein and the potential DNA binding sites. When the two are mixed together, complexes form between the protein and the DNA. DNA in these complexes is protected from cleavage produced when nucleases are added to the preparation (Figure 8-8). Consequently, analy-

---

Figure 8-7. Walking Along DNA. *(a) Preparing the radioactive probe. A short piece of DNA is shown with two adjacent, partially overlapping regions. Region I has already been cloned and is bounded by cleavage sites for restriction endonuclease X. Region II is to be cloned and is bounded by cleavage sites for restriction endonuclease R. (1) Region I DNA is inserted into a plasmid. (2) Region I–plasmid recombinant DNA is cut by endonuclease R and X to liberate the short piece of DNA (solid) that represents the overlap between regions I and II. (3) The overlapping DNA piece is separated from all other pieces by gel electrophoresis. (4) Radioactive label is enzymatically attached to the overlapping fragment. (b) Cloning the region adjacent to that already cloned. (1) Total DNA is cut with restriction endonuclease R and is spliced into plasmid DNA, producing recombinant DNA molecules of many types. (Only the particular recombinant DNA being sought is illustrated. In this DNA the region of overlap is present because endonuclease X has not been used.) (2) E. coli cells are transformed with the recombinant plasmids, and these cells are grown into colonies on agar plates. Very few of these colonies contain the adjacent region (region II). (c) Identifying the colony containing the adjacent region. The many colonies obtained in b are tested by nucleic acid hybridization using the overlapping radioactive probe prepared in a. The colony containing region II will become radioactive, and it can be identified by methods described in Appendix III (Figure AIII-4). Plasmid DNA isolated from this colony will contain region II.*

sis of nucleotide sequences that survive nuclease treatment provides insight into repressor binding sites, and this information adds to our general understanding of gene organization and regulation (see Figure 3-10).

Since cells can be manipulated to produce large quantities of certain proteins, biologists have also asked questions about what happens to the expression of related genes. In bacteria, for example, there are many related genes encoding proteins that serve as components of ribosomes. Massive overproduction of one of these proteins shuts off the production of several others. Biologists are currently studying this regulatory process to determine how cells control the production of their many components.

PERSPECTIVE

We know in general terms how DNA acts as a repository for information. By determining the nucleotide sequences of DNA molecules, we are rapidly learning exactly what that information is. But this new knowledge will not suddenly tell us how life works. Fundamental problems remain to be solved even with

Figure 8-8. Protection of DNA by a Protein. *(a) DNA fragments that have a radioactive label (asterisk) on one end of one strand are mixed with purified protein molecules. The proteins bind to the DNA fragments, covering a short region of DNA (b) Nuclease molecules are added to the protein-DNA complexes, and the DNA strands are cut. Cutting can occur anywhere except where the protein is bound to the DNA. (c) The proteins are removed by detergent and **protease** treatments, and the double-stranded molecules are converted into single-stranded ones. The nuclease treatment in b causes the DNAs to have many different lengths; only radioactive DNAs are considered because only radioactive ones are measured in d. Note that not all DNA sizes are present because the protein blocks the cutting in certain regions during the nuclease treatment in b to produce a protected region of DNA. (d) The lengths of the radioactive DNAs are analyzed by gel electrophoresis; no bands are observed that have ends occurring in the protected region.*

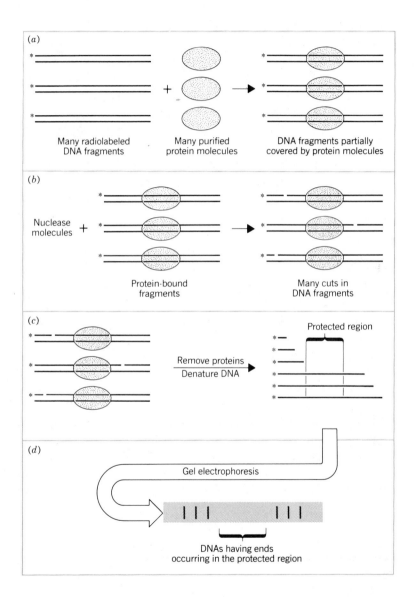

(a)

Many radiolabeled DNA fragments

Many purified protein molecules

DNA fragments partially covered by protein molecules

(b)

Nuclease molecules

Protein-bound fragments

Many cuts in DNA fragments

(c)

Remove proteins
Denature DNA

Protected region

(d)

Gel electrophoresis

DNAs having ends occurring in the protected region

simple organisms such as bacteria. For example, knowing the exact order of all 4,400,000 base pairs in *E. coli* DNA will not tell us how the timing of cell division is controlled. To understand cell division, and most other cellular processes, we must know how the products of the relevant genes work and interact.

Gene cloning may have its greatest impact on biology by helping us understand the interactions among gene products, for cloning makes it much easier to apply both biochemical and genetic techniques to a single problem. These two approaches produce different, but complementary types of information. Biochemistry yields precise information about how molecules behave in a test tube, but whether they act in the same way inside living cells is a guess. Genetics provides information about how a molecule works in a living cell, but the information is imprecise.

Genetics utilizes mutations to perturb the normal function of a gene in a living cell. As the effects of the mutations are examined, ideas develop about how the genes and their products are interacting. For example, when a mutation in a particular gene causes DNA synthesis to stop, we know that the product of that gene is somehow involved in DNA synthesis. Gene cloning allows us to purify large amounts of the gene product so that its interactions with other molecules can be studied biochemically. Then we can begin to learn exactly how DNA synthesis works. Gene cloning also allows us to purify specific genes whose products we may have already studied biochemically. We can then create a mutation in the gene and put the mutated gene into a living cell. By examining the effects of the mutation and by combining that information with biochemical data, we can begin to understand how the gene functions. One of the most successful uses of this strategy has been in the study of transposition, a subject that is briefly outlined at the end of the next chapter.

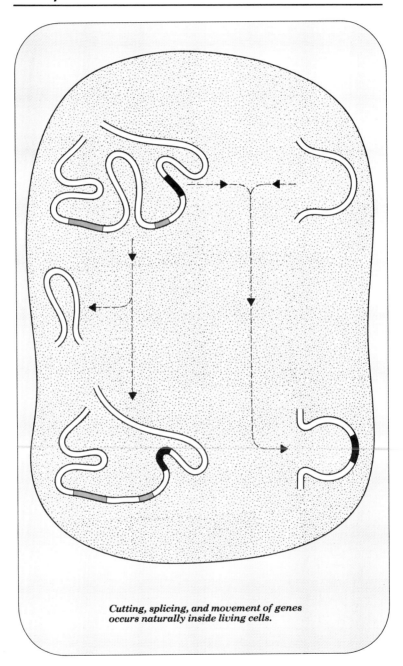

Cutting, splicing, and movement of genes
occurs naturally inside living cells.

RECENT SURPRISES

a sampling of new perspectives on the life process

overview

Gene cloning technologies have allowed us to purify large quantities of specific regions of DNA. The precise locations of many genes have been established by determining the nucleotide sequences of these regions. These studies have confirmed that the general biochemical principles of life are the same in humans and in bacteria. However, some remarkable differences exist. Unlike bacterial DNA, our DNA contains nonfunctional copies of genes. Moreover, our functional genes contain long stretches of nucleotides that code for nothing. They are spliced out of the transcript before it is translated into protein. The sequencing studies have also demonstrated that cutting, moving, and joining specific sections of DNA are normal processes that occur inside living cells.

INTRODUCTION

Gene cloning will profoundly influence two aspects of our lives: our health and our self-understanding. The two aspects are related, for increased understanding of how our bodies work leads to a greater ability to correct and prevent malfunctions. Gene cloning technologies make it possible to obtain large quantities of specific segments of DNA containing genes. Consequently, the nucleotide sequences of many of these cloned regions have been determined. Since in many cases we also know the amino acid sequences of the proteins produced from these stretches of DNA, it has been possible to delineate exactly where a number of genes are located. These cloning and sequencing studies have produced a number of surprises, four of which are briefly described in this chapter.

HEMOGLOBIN GENES AND PSEUDOGENES

The blood protein **hemoglobin** has been extensively studied for many years, and a number of statements can be made about it. First, hemoglobin is composed of four subunits, four separate protein chains called **globins**, that spontaneously associate to form the active protein. The four separate protein chains are of two types, alpha (α) and beta (β). The two types of protein differ slightly in length and in amino acid sequence and are paired in the hemoglobin molecule. Thus the predominant adult hemoglobin is generally called $\alpha_2 \beta_2$ (Figure 9-1). Second, several

Figure 9-1. Structure of Hemoglobin. *Hemoglobin is composed of two each of two types of protein subunit. In the predominant forms of adult hemoglobin, these subunits are called alpha (α) and beta (β).*

kinds of hemoglobin exist, and at different stages of life our genes instruct our blood cells to produce different globin proteins. Each kind of hemoglobin is distinguished by having different types of subunit chains. For example, the blood of young human **embryos** contains two kinds of embryonic hemoglobin, $\alpha_2 \, \varepsilon_2$ and $\xi_2 \, \varepsilon_2$. After 8 weeks of gestation the embryonic forms are gradually replaced by the **fetal** form of hemoglobin, $\alpha_2 \, \gamma_2$. Fetal hemoglobin, which predominates until about 6 months after birth, is replaced by the adult forms, $\alpha_2 \, \delta_2$ and $\alpha_2 \, \beta_2$. The third statement that can be made is that separate genes code for each of the hemoglobin subunits, ξ, α, ε, γ, δ, and β. Thus we are faced with the question of how the various genes are switched on and off during our development to produce the correct kinds of hemoglobin for each stage.

Recombinant DNA technologies and DNA sequencing studies have not yet answered how gene switching occurs, but they do allow us to make four more statements about how the genes are organized (see Figure 9-2 for schematic drawing of globin gene arrangement). First, globin genes fall into two classes. The α class includes genes α and ξ, while the β class includes genes β, γ (G and A), ε, and δ. Second, the genes in one class are located in the same region of DNA, but members of the other class are located far away on another chromosome. Third, the proteins in one class, and thus the genes that code for them, have similar structures. For example, all the protein products from the embryonic and adult α class genes are 141 amino acids long and vary only slightly in amino acid sequence. Fourth, the human genes in a class map in the same order in which they are expressed during development. For example, the β-class genes map in the order ε (embryonic), γ^G (fetal), γ^A (fetal), δ (adult), and β (adult) (Figure 9-2). Order is also preserved in the synthesis of messenger RNA from the genes. For each gene, RNA synthesis starts at the end closest to the embryonic gene (arrows, Figure 9-2), and all the messenger RNAs are made from the same strand of DNA (the two DNA strands are complementary, not identical, and they would not code for the same proteins). This ordering of genes and direction of transcription

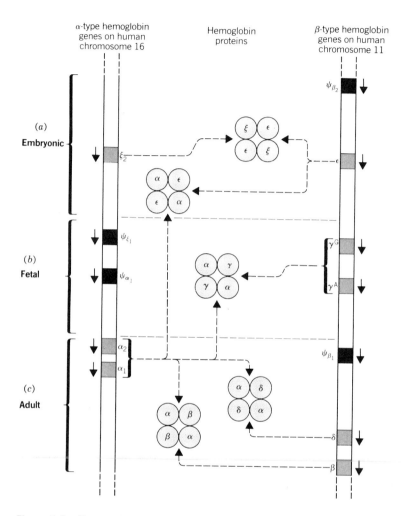

Figure 9-2. Human Globin Genes and Their Protein Products. *Hemoglobin (depicted as four circles) is composed of two types of protein, and the composition varies with the stage of development through differential activation of globin genes. The embryonic forms (a) predominate until 8 weeks of gestation, the fetal form (b) to 6 months after birth, and the adult forms (c) from 6 months on. Heavy arrows indicate the direction of transcription, solid areas represent* **pseudogenes** *(nonfunctional genes; see text). The β-type genes are scattered over a stretch of 52,000 nucleotide pairs, whereas the α-type genes fall within a 36,000 nucleotide-pair region.*

is too consistent to be accidental. Is it a clue to understanding developmental switching of genes?

Sequencing studies of the globin gene regions also uncovered several other short stretches of DNA that look remarkably like bona fide globin genes. Careful analysis of the nucleotide arrangements showed that these "genes" (solid regions (ψ) in Figure 9-2) are incapable of producing functional globin proteins—they are full of mutations and abnormalities. Premature stop codons, frameshift mutations, abnormal RNA polymerase binding sites (promoters), faulty initiation codons, and large internal deletions ensure that no functional globin proteins can come from these genelike regions called **pseudogenes**. The surprising discovery of pseudogenes confronts us with a new set of questions. Did pseudogenes arise from the duplication of a preexisting gene? Did all the modern globin genes arise by duplication of a primitive ancestor? How does gene duplication occur?

The hemoglobin system is complicated, and, as in all complicated biological systems, there are many steps at which things can go wrong and produce serious diseases. One class of disease arises from nucleotide substitutions in the genes, causing amino acid changes in the globin proteins. Of the 300 or so mutations that have been identified, the most widely known is the mutation in the β-globin gene that causes sickle-cell anemia (see Figure 4-5). Another group of hemoglobin disorders, called the thalassemias, comes from a deficiency of one specific type of globin protein. Still other problems arise when the normal switch from one form of hemoglobin to another fails to occur.

As we begin to understand the molecular basis for some of these diseases, the prospect for cures grows. Consider, for example, the case of sickle-cell anemia. The disease could be cured by replacing the faulty β-globin protein. The sickle-cell patient already has three other forms of β-type globin, ε, γ, and δ. If we knew how the γ–β switch works, we might be able to switch off the sickle-cell β-globin gene and switch on the fetal γ-globin gene, thereby indirectly curing the disease. The blood of the patient would then contain normal fetal hemoglobin rather than the defective adult form. The fetal form might not work quite as well as the normal adult form; however, adults who naturally have fetal hemoglobin rather than adult hemoglobin appear to be healthy. Attempts are being made to put this particular scenario into practice, illustrating the close relationship between basic biological research and medicine.

EXONS, INTRONS, AND RNA SPLICING

Once genes were purified, it was a straightforward process to determine the nucleotide sequences of the genes and of the DNA surrounding them. In higher organisms many cases were soon found in which the nucleotide sequence of the gene did not match that predicted from the amino acid sequence of the protein: the coding regions of a gene were often observed to be interrupted by stretches of nucleotides that did not code for animo acids found in the protein. The organization of the human β-globin gene is shown in Figure 9-3 as a gene containing three coding regions, called **exons**, interrupted by two noncoding regions, or **introns**. Cases of more than 50 introns scattered through a single gene have been reported. The RNA molecules transcribed from genes containing introns are longer than the messenger RNAs that subsequently produce the protein specified by the gene. Thus we hypothesize that cells have

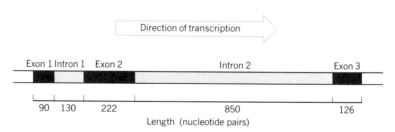

Figure 9-3.   Intervening Sequences in the β-Globin Gene of Humans. *The exons code for the amino acids in the protein.*

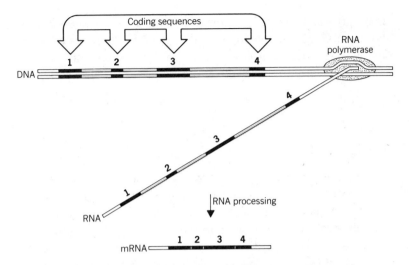

Figure 9-4. Arrangement of Sequences in a Gene from A Higher Organism. *The regions of DNA coding for amino acids in the protein product are interspersed with noncoding regions, which are processed out before mature messenger RNA is formed. Coding sequences 1 through 4 are part of a single gene. Note that in high organisms ribosomes do not bind to messenger RNA before it is released from the DNA (see Figure 3-7 for comparison with bacteria).*

mechanisms for removing the introns. In the process, information in the exons is spliced together to form mature messenger RNA (Figure 9-4).

How the splicing reaction occurs is the subject of current research. Bacterial genes do not contain introns; thus, in bacteria, splicing is not a part of messenger RNA processing. Why should there be such a dramatic difference between bacteria and higher cells? Perhaps in an evolutionary sense RNA splicing developed after bacteria and higher cells had become separate types of life forms. RNA splicing may provide higher cells with many gene control options lacking in bacteria. These options may have made it possible to develop multicellular bodies with complicated parts. Alternatively, splicing may have characterized the earliest forms of life, and through evolution bacteria may have lost the ability to splice RNA. Although too little is known about RNA splicing to make general statements concerning its function, examples of its importance are being found. A particularly striking example is discussed in the next section.

ANTIBODIES

Antibodies are proteins that recognize and bind to foreign substances, that is, substances not normally found in our bodies. As such, antibodies form part of our immune system, the elaborate network of molecules and cells that protects us from many types of disease. When an antibody attaches to a foreign substance, which is called an **antigen**, a number of steps are activated in the immune system that result in destruction or expulsion of the antigen. Millions of antigens can be recognized by antibodies. Since each different antigen is recognized by a different antibody, our bodies must be able to produce antibodies of millions of different types.

Each antibody is composed of four protein chains, two identical **heavy chains** and two identical **light chains**. The chains are folded and connected to form a "T" as shown in Figure 9-5. Comparison of amino acid sequences from many different antibodies has revealed several interesting features. First, antibodies can be grouped into classes based on the amino acid sequences and properties of the heavy chains. Second, within a class there are sections of the protein chains that are identical from one antibody to the next. These sections are called **constant regions**, and they determine the behavior of the antibody in our bodies. For example, heavy chain antibodies with one type of constant region circulate in the blood, those with another type attach to the surface of the cell that produced them, and still others bind to specific cells that release histamines. The third point is that each light chain and each heavy chain have regions of amino acids that are unique to that antibody. These regions are called **variable regions**, and it is this

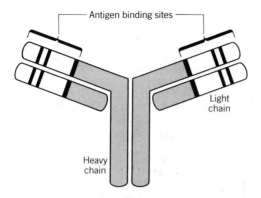

Figure 9-5. Schematic Diagram of an Antibody Molecule. *Two heavy chains pair with each other and with two light chains to form the active antibody. The amino acid sequences are divided into constant regions (shaded), variable regions (open), and hypervariable regions (solid). Two antigen binding sites are present, one in the variable region of each arm.*

part of the antibody that binds to foreign substances like viruses and bacteria. Since the shape and structure of a protein are dramatically affected by small changes in the sequence of amino acids, the slight differences in amino acids found in the variable regions result in millions of different antibodies, each able to recognize a particular antigen.

For several decades biologists were puzzled about how so many different antibodies could be produced. Could there be millions of genes, one for each protein chain of every antibody? Calculations suggested that we might not have enough DNA to code for all the antibodies and still have enough genes to run the chemistry of our cells. DNA sequencing studies now reveal that most of our cells do *not* have a complete set of antibody genes. Instead, they have bits and pieces that can be combined in a number of different ways, thus producing millions of distinct antibodies from a small amount of genetic information. The rearrangements occur inside special blood cells called **B lymphocytes,** and these cells are responsible for making antibodies.

By comparing the nucleotide sequences in DNA from embryonic cells with those in DNA from antibody-producing cells, it has been possible to develop a general idea about how gene shuffling results in antibody chains. In the case of light chains (Figure 9-6), the embryo contains several hundred variable region genes (V) widely separated from five short, joining genes (J). DNA breakage and rejoining occurs so that one of the V genes is placed next to one of the J genes. RNA polymerase transcribes this region and continues until it also transcribes a

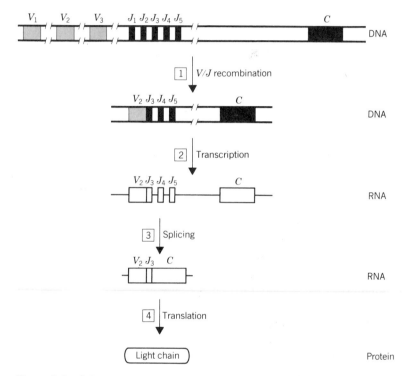

Figure 9-6. Schematic Representation of the Formation of an Antibody Light Chain. *(1) One of the approximately 150 variable genes V recombines with one of the 5 joining genes J. In the example V_2 is moved so it becomes adjacent to J_3. (2) RNA is synthesized from this DNA to produce a primary transcript. (3) Splicing occurs to remove all the RNA between J_3 and the constant gene C, producing mature messenger RNA. (4) This messenger RNA is translated into the antibody light chain. Discontinuities in the DNA indicate large distances between the genes.*

constant region gene (C). This long RNA molecule is then spliced to remove the sequence between the *V/J* region and the C region, producing mature messenger RNA. The messenger RNA is then translated into an antibody light chain. Since any one of perhaps 150 *V* genes can join to any of 5 *J* genes, roughly 750 combinations (150 × 5) can occur. Moreover, the joining sites are not precisely located; thus, the actual number of possible combinations is probably closer to 7500.

The same principles apply for heavy chain formation. However, more elements are included in creating heavy chain diversity (Figure 9-7). In humans there are about 80 *V* (variable) genes, 50 *D* (diversity) genes, and 6 *J* (joining) genes. Thus there are about 24,000 combinations (80 × 50 × 6) that can form. Flexibility in the *V/D* and the *D/J* junctions probably adds 100 more ways to combine the genes, so the total number of heavy chain combinations is about 2.4 million (24,000 × 100). The total number of antibody combinations is the product of the light chain and heavy chain combinations, or 18 billion (7500 × 2.4 million). Thus enormous diversity can be produced by about 300 embryonic DNA segments.

As mentioned earlier, there are several types of heavy chain, each with a different constant region. These constant regions determine how the antibody will behave in the body. Constant region genes are arranged downstream from the *J* region, and by selective splicing and additional recombination it is possible to put the identical *V/D/J* region onto five different constant regions. Consequently, our bodies can contain a number of antibody types that recognize the same foreign substance but perform different functions to combat it.

In summary, we appear to combat foreign substances in the following way. Our blood has millions of B lymphocytes circulating in it, and each B lymphocyte undergoes a slightly different DNA rearrangement. Thus each one produces antibody molecules that are slightly different from those produced by other B lymphocytes. Each cell produces several classes of antibodies that have different heavy chain constant regions. The function of one of these classes is to reside on the surface of the B

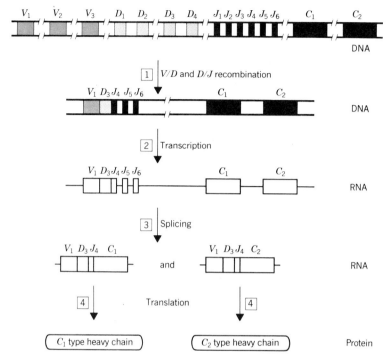

Figure 9-7. Schematic Representation of the Formation of Antibody Heavy Chains. *(1) One of 80 V regions joins with one of about 50 D regions and one of 6 J regious to form a recombined DNA molecule in a cell called a B lymphocyte. (2) A primary transcript is made that contains two different C regions. (3) By differential splicing, two types of heavy chain messenger RNA can be made. (4) When the messenger RNAs are translated, they produce two types of heavy chain protein. Since the $V_1/D_3/J_4$ regions are the same for both, the two heavy chain proteins will have identical antigen binding sites. Discontinuities in the DNA indicate large distances between the genes.*

lymphocyte that produced it. There the antibody acts as a sentry, waiting to intercept a foreign substance, an antigen. When the antibody on the cell encounters an antigen to which it can bind, a complex is formed between the antigen and the antibody. The complex then triggers that particular lymphocyte to multiply and to produce additional antibody molecules. All the antibodies from a particular cell line have identical variable regions, so they all recognize the same antigen. The constant regions vary, however, so we end up with antibodies that can function in different ways to rid our bodies of the antigen. Moreover, some antibodies will recognize one part of a foreign substance while others may recognize different parts. Thus a complex structure like a virus could stimulate a number of different B lymphocytes to produce antibodies, increasing our chances of fighting off infection.

TRANSPOSITION

Gene cloning technologies allow us to insert small, discrete fragments of DNA into specific places in other DNA molecules. Nature also has this ability. Scattered among the genes of living cells are small, discrete sequences of nucleotides that can duplicate themselves so that the new copy is able to hop to another DNA molecule (Figure 9-8). These discrete nucleotide sequences are called **transposons**, and the process in which a

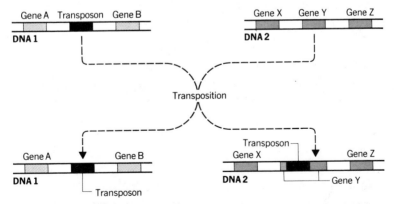

Figure 9-8. Transposition. *A specific region of DNA 1 called a transposon (solid region) duplicates itself and inserts a copy into DNA molecule 2. In this example the transposon inserts in the middle of gene Y, splitting the gene into two parts.*

transposon moves to another location in DNA is called **transposition**. Transposition can have several consequences, depending on where the new copy of the transposon becomes inserted. For example, when a transposon inserts into the coding region of a gene, the information in the gene is interrupted (gene Y, Figure 9-8), and the gene may no longer produce a functional protein. Cases have also been observed in which transposition activates a gene by insertion near the gene.

Transposons have been found in a wide variety of organisms, and it is likely that all organisms contain them. Thus far transposons from different organisms are known to share three common features. First, transposons are always discrete sections of DNA; the junctions between a transposon and the DNA in which it is inserted are precisely defined. Second, transposons usually contain nucleotide sequences encoding one or more products required for movement of the transposon from one site to another. In the best studied cases these products turn out to be proteins. Thus transposons contain genes responsible for their own movement. Third, each end of a transposon contains nucleotide sequences that probably serve as recognition sites for factors involved in movement of the transposon. These sequences are repeats of each other; in most cases the repeats are inverted. A general scheme of transposon structure appears in Figure 9-9.

Although the molecular details of transposition vary from one type of transposon to another, a description of one called Tn3 is sufficient to provide an appreciation for the process. Tn3 is found in bacterial cells, and it contains three genes, A, R, and bla (Figure 9-10). In addition, Tn3 has a stretch of 38 base pairs at its left end that is repeated at its right end in an inverted orientation. The bla gene encodes a protein that destroys ampicillin (penicillin); thus any cell containing Tn3 is resistant to ampicillin. This feature helps biologists determine whether a cell contains Tn3. The A gene encodes a protein called **transposase**, a protein responsible for Tn3 movement. If small regions of the A gene are experimentally removed, Tn3 can no longer move. The R gene codes for a repressor that binds to the A gene and

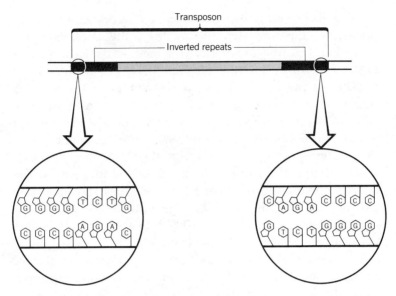

Figure 9-9.   General Structure of a Transposon. *Transposons contain repeated nucleotide sequences at each end. The repeated sequences are generally in an inverted orientation. Genes involved in transposon movement lie between or within the repeats.*

prevents it from making transposase. Thus the repressor keeps transposition from occurring very often. When the *R* gene is damaged by mutation, the frequency of the first step in Tn3 transposition (see below) increases greatly.

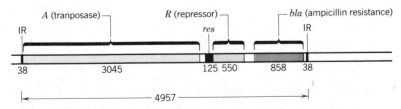

Figure 9-10.   Arrangement of Genes in Tn3. *Tn3 contains three genes, A, R, and bla, located between the 38 base-pair inverted repeats (IR). The repressor protein binds to Tn3 DNA at region res. Numbers indicate nucleotide pairs in each region.*

It is likely that transposition of Tn3 occurs as a two-step process. In the first step (Figure 9-11) the DNA molecule containing Tn3 (the donor DNA) somehow binds to a DNA molecule lacking Tn3 (the recipient DNA). This process is mediated by the transposase protein. Then the Tn3 sequence is duplicated to produce a structure called a **cointegrate**. The cointegrate contains two copies of Tn3, one at each junction between the donor and recipient DNAs. After the cointegrate is formed, transposase dissociates from the DNA.

In the second step the cointegrate is split into two circles, and both donor and recipient DNA molecules end up with a copy of Tn3. The product of the *R* gene, the repressor, plays a key role in this process (Figure 9-12). It binds to the cointegrate DNA, and it probably twists the DNA to align the two copies of Tn3. Breaks occur in the two copies of Tn3 at the *res* sites (Figure 9-10), one double-stranded DNA in effect crosses over the other,

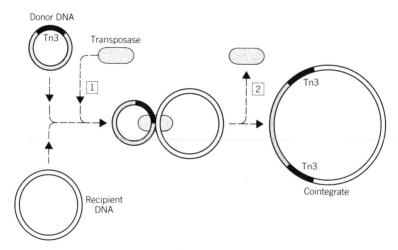

Figure 9-11. Scheme for Formation of Cointegrates by Tn3. *(1) Transpos-ase mediates the joining of a DNA molecule containing Tn3 (donor DNA) with a DNA lacking Tn3 (recipient DNA). (2) Tn3 is replicated (duplicated by DNA polymerase), and in the process the donor and recipient DNAs are joined to form a larger circle (cointegrate). The transposase leaves the DNA and is presumably free to initiate a new round of transposition.*

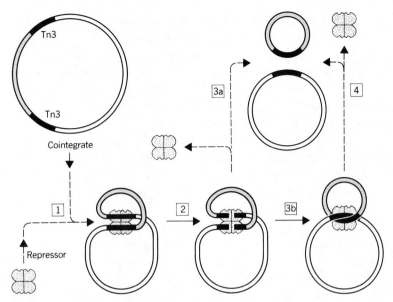

Figure 9-12.   Recombination Between Two Tn3 Transposons in a Cointegrate. *(1) The repressor protein binds to the cointegrate, probably twisting the DNA molecule so that the two copies of Tn3 (solid) align. (2) A break occurs within the* res *site (see Figure 9-10) of each Tn3. (3) One DNA crosses over the other, and the breaks are sealed. This produces two rings, one of which is the donor (shaded) and the other is the recipient (open). Both contain a copy of Tn3. Recombination has two possible outcomes. (3a) Two separate rings are produced (3b) Two interlocked rings arise. In both cases the repressor dissociates from the rings (3a, 4). (4) An enzyme called a topoisomerase will separate the interlocked rings.*

and the broken ends are joined so that a part of one Tn3 is linked to the remainder of the other. This process of DNA strand exchange is called **recombination**; genetic information in the two copies of Tn3 has been exchanged. In this case recombination also creates two DNA circles, each with a copy of Tn3. Thus the repressor plays two roles in transposition of Tn3: first, it keeps the frequency of transposition low by repressing the *A* gene, and second, it is required to align, break, and rejoin cointegrate DNA. Another interesting feature is that the repressor also represses the *R* gene—it represses its own production and keeps the number of recombination events very low. Thus transposition is a very tightly regulated process.

Transposons are among the most exciting new tools for genetic research, and it is probably fitting to close this chapter with a few speculations. First, transposons can be used as cloning vehicles. Genes can be inserted into a transposon already in a plasmid by gene splicing technologies, and many copies of the engineered transposon can be obtained by putting it inside bacterial cells. The plasmids, along with the transposon and the gene it carries, can be purified from the bacteria by methods described in Chapter 6. They can then be injected into animals, where the transposon, with the gene it carries, will duplicate and hop into the animal chromosomal DNA. Such a manipulation has already been performed to change the color of eyes in fruit flies. Second, it is possible that some genetic diseases are caused when transposons inactivate certain genes. If we had a way to remove transposons, we might be able to reactivate the gene and thus cure the disease. Conversely, some diseases may turn out to arise from overactive genes. Perhaps it will become possible to specifically inactivate these genes by transposons and cure the disease. One final observation can be added to these speculations: some viruses that cause tumors have their nucleotide sequences organized much like those of transposons. Perhaps understanding transposition will lead to a cure for some types of cancer.

PERSPECTIVE

It has been slightly more than 10 years since methods for cloning genes were developed. Now almost every molecular biology laboratory is using them, and even undergraduate students are being taught how to determine nucleotide sequences of DNA. We can expect more scientific surprises, requiring frequent revision of books on molecular biology. We can also expect basic research to produce new ways to combat cancer and other diseases. World food supplies should also improve as plants and domestic animals are engineered to thrive in areas of the world that are currently unproductive. Thus we will all be affected by gene cloning.

You have now completed a brief, nine-chapter tour of molecular biology. I hope the tour has aroused your curosity about life and has given you a sense of the excitement taking place in biology. Additional readings for those wishing to continue the study of genes are listed below.

additional reading

Anderson, W. F., and Diacumakos, E. G. (1981) Genetic Engineering in Mammalian Cells. *Scientific American* (July 1982), 106–121.

Strategies are described for using bacteria and recombinant DNA techniques to engineer mammalian cells.

Bishop, J. M. (1982) Oncogenes. *Scientific American* (March 1982), 80–92.

Oncogens cause cancer. They were first reported in viruses, but they have also been found in normal cells. Abnormal expression of these genes can lead to cancerous growth.

Brierley, C. L. (1982) Microbiological Mining. *Scientific American* (August 1982), 44–53.

A type of bacterium called *Thiobacillus* may become very useful in leaching copper from low-grade ore.

Capra, J. D., and Edmundson, A. B. (1977) The Antibody Combining Site. *Scientific American* (January 1977), 50–59.

Antibody structure is discussed, and the biochemical details of antigen–antibody binding are described.

Chambon, P. (1981) Split Genes. *Scientific American* (May 1981), 60–71.

Some of the experimental evidence leading to the concept of introns and exons is presented.

Chilton, M.-D. (1983) A Vector for Introducing New Genes into Plants. *Scientific American* (June 1983), 51–59.

Some plant tumors are caused by a bacterium that may soon be used to introduce genes into plants.

Freifelder, D., Editor (1978) *Recombinant DNA*. W. H. Freeman & Company, San Francisco. 147 pp.

A collection of articles from *Scientific American* that provides background material on the development of recombinant DNA technology.

Geever, R. F., Wilson, L. B., Nallaseth, F. S., Milner, P. F., Bittner, M., and Wilson, J. T. (1981) Direct Identification of Sickle-Cell Anemia by Blot Hybridization. *Proceedings of the National Academy of Sciences (U.S.)*, **78**, 5081–5085.

This research paper describes how sickle-cell anemia can be diagnosed.

Gilbert, W., and Villa-Komaroff, L. (1980) Useful Proteins from Recombinant Bacteria. *Scientific American* (April 1980), 74–94.

Recombinant DNA techniques are described with special reference to the procedures used to clone an insulin gene.

Grobstein, C. (1979) *A Double Image of the Double Helix*. W. H. Freeman & Company, San Francisco. 177 pp.

During the mid-1970s recombinant DNA technologies led to public controversies. This book discusses the controversies and provides a copy of the NIH guidelines that regulate recombinant DNA research.

Jonathon, P., Butler, G., and Klug, A. (1978) The Assembly of a Virus. *Scientific American* (November 1978), 62–69.

This description of the assembly of Tobacco Mosaic Virus provides an introduction into the details of virus structure.

Kornberg, R. D., and Klug, A. (1981) The Nucleosome. *Scientific American* (February 1981), 52–64.

Higher cells package their DNA by wrapping it around ball-like structures made of protein.

Lake, J. A. (1981) The Ribosome. *Scientific American* (August 1981), 84–97.

A three-dimensional model is presented in this description of how proteins are made.

Leder, P. (1982) The Genetics of Antibody Diversity. *Scientific American* (May 1982), 102–115.

The shuffling of segments of DNA and RNA that occurs during the formation of antibody genes is considered.

Lewin, B. (1983) *Genes*. John Wiley & Sons, New York. 715 pp.

This thorough molecular genetics textbook covers topics such as the nature of genetic information, the synthesis of proteins and templates, the control of gene expression, the organization of information in DNA, the packaging of DNA, and the movement of genes.

Maniatis, T., Hardison, R. C., Lacy, E., Lauer, J., O'Connell, C., Quon, D., Sim, G., and Efstratiadis, A. (1978) The Isolation of Structural Genes from Libraries of Eucaryotic DNA. *Cell*, **15**, 687–701.

This research paper describes the cloning of rabbit hemoglobin genes.

Pestka, S. (1983) The Purification and Manufacture of Human Interferons. *Scientific American* (August 1983), 36–43.

Interferons are antiviral proteins that are now being produced by genetic engineering.

Ptashne, M., Johnson, A. D., and Pabo, C. O. (1982) A Genetic Switch in a Bacteria Virus. *Scientific American* (November 1982), 128–140.

When bacteriophage lambda infects a cell, it can either kill its host or coexist with it. This article describes how the virus makes the choice.

Simons, K., Garoff, H., and Helenius, A. (1982) How an Animal Virus Gets into and out of Its Host Cell. *Scientific American* (February 1982), 58–66.

The virus causes the host cell to manufacture new virus particles, including a protective membrane that originates from the host cell membrane.

Wang, J. C. (1982) DNA Topoisomerases. *Scientific American* (July 1982), 94–109.

A class of enzyme is described that is able to convert rings of DNA from one topological form to another. These enzymes are probably important in many aspects of chromosome function.

Watson, J. D. (1976) *The Molecular Biology of the Gene*, 3rd ed. W. A. Benjamin, Menlo Park, Calif. 739 pp.

> This very popular textbook on molecular biology makes the chemical aspects of biology accessible to college undergraduates.

Watson, J. D. (1980) In *The Double Helix* (Gunther Stent, Ed.) W. W. Norton & Company, New York. 298 pp.

> Watson's account of the discovery of DNA structure is placed in historical perspective by Gunther Stent, Francis Crick, Linus Pauling, and Aaron Klug. Stent has also collected background materials, which include reproductions of original scientific papers and reviews of Watson's story by well-known scientists.

Watson, J. D., and Tooze, J. (1981) *The DNA Story*. W. H. Freeman & Company, San Francisco. 605 pp.

> Included in this documentary history of gene cloning are reproductions of press clippings and personal letters written by scientists involved. The section titled "Scientific Background" provides a historical introduction to gene cloning.

appendix I

ATTENUATION

a type of gene regulation

Attenuation is a mechanism for controlling gene expression in which the synthesis of messenger is halted after only a short portion of the messenger has been made. This process has been most extensively studied with genes involved in making amino acids. Amino acids are essential for the health of the cell as building blocks for proteins, so they need to be kept in constant supply. However, their production costs the cell a considerable amount of energy. As a result, mechanisms have evolved that carefully control amino acid production and maintain the right balance of the 20 different amino acids. Attenuation is one of these mechanisms.

In the attenuation of the **tryptophan** genes, RNA polymerase begins making RNA some distance from the beginning of the first gene in the message, thus creating a **leader** RNA. This leader RNA contains a coding region for a short leader protein, and some of the codons in this region specify that tryptophan is to be inserted into the leader protein. Within this leader region is also a variably active stop signal called an attenuator. When tryptophan is abundant, ribosomes are able to translate the leader protein, and as a consequence, the attenuator region of the RNA folds in such a way that RNA polymerase is stopped. Only the short leader message is made. When tryptophan is scarce, the ribosomes stall when they come to the tryptophan codons in the message for the leader protein. The stalling prevents the attenuator from halting RNA polymerase. The enzyme continues down the DNA, making the entire message,

169

which contains the genes for proteins involved in tryptophan production. Each of these genes has its own start site on the messenger RNA for ribosome binding and protein production. Thus, the level of tryptophan in the cell controls the production of more tryptophan.

appendix II

TRANSMISSIBLE PLASMIDS

moving genes

Transmissible plasmids, like all infectious agents, are self-serving, and they contain a number of genes whose protein products enable the plasmid to go from one bacterial cell to another. The process of plasmid movement is called **conjugation**. The best studied case of conjugation involves a plasmid of *E. coli* called **F** or **fertility factor**. The plasmid DNA contains genes coding for proteins responsible for the formation of long filamentlike structures that protrude from the outside of the bacterial cell. Such a protrusion is called a **pilus; pili** are thought to be important in attaching or attracting plasmid-containing cells to plasmid-deficient ones. When the two types of cells get close enough, they mate. A new copy of the plasmid DNA is made and is transferred from the plasmid-containing cell to the plasmid-deficient one. Thus both cells of the mating pair end up infected with a plasmid (Figure AII-1). On rare occasions the plasmid DNA and the bacterial chromosomal DNA can join to form a giant circle. When this happens, genes in the bacterial chromosome can be transferred from one cell to another by the plasmid mating process. In a sense, bacterial conjugation is a primitive form of sex.

In general, transmissible plasmids are not considered to be good cloning vehicles because they could conceivably lead to accidents. For example, most laboratory strains of *E. coli* are useful as tools because they contain a number of known genetic defects. As a result of the defects, these weakened organisms

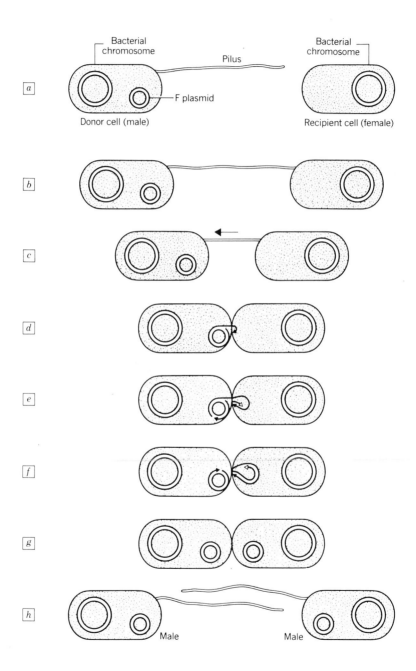

are not expected to survive well outside the laboratory, and in general, the laboratory strains seem unlikely to cause infections. However, if genes were to be cloned into laboratory bacteria by using transmissible plasmids, and if the "engineered" bacteria containing the plasmids and cloned genes were to accidentally escape from the laboratory, through the process of conjugation the plasmids and the cloned genes could conceivably be transferred into a healthy, wild strain of the bacterium. The wild strain might have a greater chance than the weak, laboratory strain to enter human digestive tracts and thus place an unwanted gene where it could do harm. Fortunately there are a large number of *non*transmissible plasmids of varying sizes and degrees of complexity, so it is unnecessary to use transmissible plasmids for gene cloning.

Figure AII-1. Bacterial Conjugation. *(a) An E. coli cell containing an F plasmid forms long, flexible tubular structures (pili) on its surface (1–3 per cell). Each pilus is composed of proteins encoded by genes located on the F plasmid. (b) One pilus binds to an E. coli cell that lacks an F plasmid. (c) The pilus retracts, pulling the two cells close to each other. (d) A break occurs in one of the DNA strands of the F plasmid; one of the ends of the broken strand rolls off the circular strand and passes into the recipient cell. (e, f) Soon after the single strand has reached the interior of the recipient cell, DNA polymerase (not shown) begins to make a complementary strand. At about the same time, a new copy of the transferred strand is made in the donor cell. (g) Both donor and recipient cells now have a complete copy of the F plasmid. (h) The two cells separate. Each contains a copy of the F plasmid. The recipient cells forms pili. Now both can act as donor cells.*

appendix III

CLONING WITH PLASMIDS

a general method

A general plasmid cloning strategy is described (Figure AIII-1) to illustrate some of the unique features of plasmids. After human DNA fragments have been spliced into plasmid DNA molecules (Chapter 5), the spliced DNA mixture is added to *E. coli* cells in a flask, and some of the bacteria take up the DNA by a process called transformation. The bacteria continue to grow and divide, and soon there are billions of bacteria in the flask. At this stage the problem is to locate the few bacterial cells that contain the cloned human genes. The *E. coli* cells in the flask can be divided into four classes.

1. *E. coli* that failed to take up any plasmid DNA.
2. *E. coli* that took up plasmid DNA without any human DNA spliced in.
3. *E. coli* that took up plasmid DNA with human DNA spliced in but not the particular human gene being sought.
4. *E. coli* that took up plasmid DNA into which the desired human gene was spliced.

The fourth category is the important one, and members of this category are generally very rare, perhaps one in a billion.

Two tricks are used to increase the odds for finding plasmids with human genes. First, the plasmid chosen as a cloning

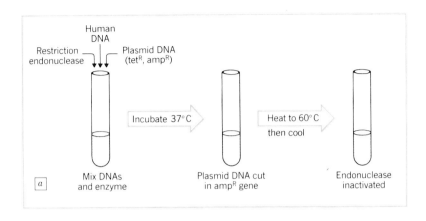

Human DNA

Restriction endonuclease — Plasmid DNA (tet^R, amp^R)

Incubate 37°C

Heat to 60°C then cool

Mix DNAs and enzyme

Plasmid DNA cut in amp^R gene

Endonuclease inactivated

a

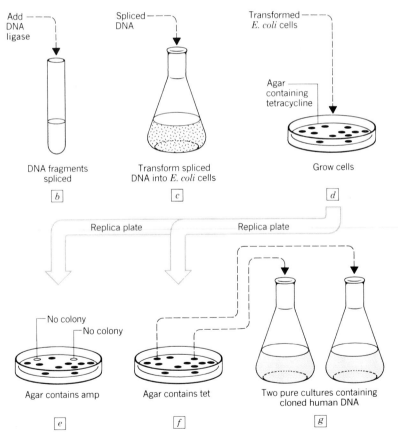

Add DNA ligase

Spliced DNA

Transformed *E. coli* cells

Agar containing tetracycline

DNA fragments spliced

Transform spliced DNA into *E. coli* cells

Grow cells

b

c

d

Replica plate

Replica plate

No colony

No colony

Agar contains amp

Agar contains tet

Two pure cultures containing cloned human DNA

e

f

g

vehicle contains a gene for resistance to **tetracycline**, a drug that kills bacteria and is used to cure diseases such as bacterial pneumonia. Thus the antibiotic could simply be added to the flask to kill cells that are not protected by that plasmid, and all the cells in category 1 would be eliminated. Alternatively, all the cells can be spread onto an agar plate that contains tetracycline. Then only the cells containing a plasmid having a gene for tetracycline resistance will grow and form colonies; cells without plasmids will die.

The second trick distinguishes cells containing plasmids spliced with human DNA from those that do NOT have human DNA in them (Figure AIII-1). To execute this trick we must use a cloning vehicle, a plasmid, with two genes for antibiotic resistance. Often one gene is for tetracycline resistance (tet^R) and the other for ampicillin (penicillin) resistance (amp^R). Since restriction endonucleases cut in very specific locations, an endonuclease can be found that cuts the plasmid DNA only inside the ampicillin-resistance gene (Figure AIII-2). Consequently, whenever human DNA is attached to this plasmid, it will be spliced into the middle of the ampicillin-resistance gene, for that is where the ends of the DNA occur. The large chunk of human DNA inserted into this gene will cause the gene to be inactive. Cells containing plasmids into which human DNA is inserted into the ampicillin-resistance gene will be resistant only to

---

Figure AIII-1. Procedure for Obtaining Pure Cultures Containing Cloned Human DNA. *(a) Plasmid DNA (resistant to tetracycline and ampicillin) and human DNA are mixed, cut with a restriction endonuclease, and incubated at 37° C. A nuclease is chosen that cuts the plasmid once. That cut occurs in the ampicillin-resistance gene (amp^R). After the DNAs are cut, the endonuclease is inactivated by heating the mixture to 60° C. (b) DNA ligase is added to splice the DNA fragments. The human DNA is spliced into the middle of the amp^R gene, inactivating it (see Figure AIII-2). (c) E. coli cells are transformed with the spliced DNA. (d) Plasmid-containing cells are selected by growth on agar containing tetracycline. Cells with plasmids spliced to human DNA are identified by screening on ampicillin-containing agar (e) [these cells grow only on tetracycline (f)]. Pure cultures (g), which contain cloned genes, result.*

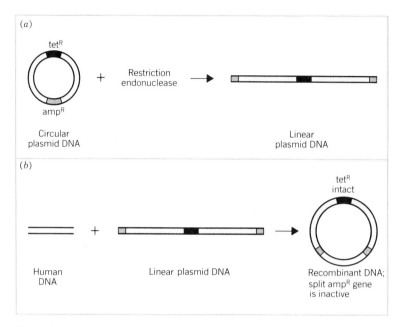

Figure AIII-2. Inactivating a Gene. *(a) Plasmid DNA containing genes for resistance to ampicillin and to tetracycline is cut with a restriction endonuclease, dividing the amp^R gene into two parts. (b) When human DNA is spliced into the plasmid, the two parts of the amp^R gene remain separated; hence the gene is inactive.*

tetracycline. On the other hand, cells containing a plasmid having no human DNA will be resistant to both drugs. Thus all the colonies that formed on the tetracycline-containing agar plate (Figure AIII-1*d*) must be tested to find ones that *fail* to grow on ampicillin-containing agar.

A piece of sterile velvet is carefully placed on the surface of the tetracycline-containing agar plate so it touches the bacterial colonies. Some of the cells from each colony will stick to the velvet. The velvet is then removed and set onto a clean ampicillin-containing agar plate. Cells from each colony will come off the velvet and stick to the agar of the clean plate. There the cells will grow into colonies if they are resistant to ampicillin. This technique is called **replica plating**, and it distributes cells onto the second plate in a pattern identical to the distribution of the colonies on the first plate. This procedure is outlined in Figure AIII-1; two colonies are present in Figure AIII-1*f* that are absent in Figure AIII-1*e*. These colonies contain human DNA spliced into the plasmid. Each of the two colonies is touched with a sterile wire and portions are transferred into separate flasks containing sterile broth (Figure AIII-1*g*). The cells multiply, producing a pure culture; all the cells in the culture are identical and contain a discrete piece of human DNA cloned into a plasmid. Thus, the field has been narrowed to a collection of pure cultures having plasmids with cloned human DNA.

But at this stage one doesn't know which gene or genes are contained in any particular culture. In fact, the chance of obtaining one that has the gene of interest is very low. Consequently, the plating procedure must be repeated many times to produce a large collection of cultures. The replica plating technique makes it possible to quickly screen thousands of colonies for growth in different drugs by simply looking for differences in the distribution patterns of colonies on agar plates. Once several thousand have been collected, we can proceed to the next trick—identifying bacterial clones that have the particular gene we seek.

First, a small sample of cells from each culture containing a human DNA fragment (Figure AIII-1*g*) is spotted on an agar plate to produce a gridlike arrangement. The cells grow into colonies on the agar (Figure AIII-3). After the cultures have grown, a piece of paper is placed on the agar and then removed (sometimes the cells are grown directly on filter paper placed on the agar surface). Bacterial cells stick to the paper, forming the same grid pattern they had held on the agar. The paper is placed in a dilute **sodium hydroxide** (lye or caustic soda) solution.

The sodium hydroxide breaks the cells, and some of the cell debris plus the cellular DNA stick tightly to the paper. The sodium hydroxide also causes the DNA to become single stranded. It is important to note that the DNA is arranged in spots on the paper, and the gene cloner knows which bacterial culture each spot corresponds to. Next, the sodium hydroxide is neutralized with acid, and the paper is slipped into a dish

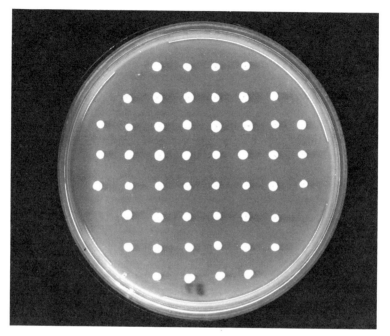

Figure AIII-3. Agar Plate Bacterial Colonies Arranged in a Regular Grid Pattern.

containing the radioactive probe, the complementary DNA made from messenger RNA from the gene being sought.

Both the radioactive probe DNA and the cellular DNA attached to the paper are single stranded, ready to form base pairs with any DNA they come in contact with. But the probe will form base pairs with paper-bound DNA *only* if the paper-bound DNA contains the gene being sought (see Figure 4-10). When this happens, the radioactive complementary DNA will be indirectly bound to the paper, and the location of the radioactivity will identify the bacterial colony which contains the human gene being sought.

To determine where the radioactivity is located, the paper is removed from the dish and washed thoroughly to remove any radioactive probe that is not base paired with paper-bound DNA. X-Ray film is then placed next to the paper (Figure AIII-4*a*). Wherever radioactive probe is base-paired to paper-bound

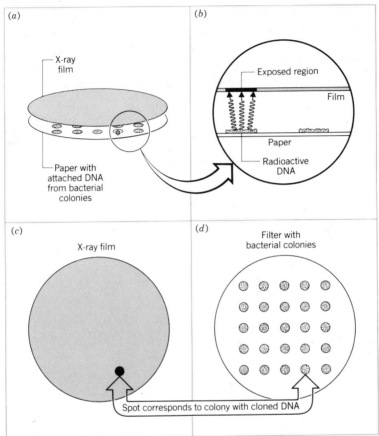

Figure AIII-4. Detection of Bacterial Colonies Containing a Particular Human Gene. *Bacterial cultures are grown on paper placed on the surface of an agar plate (nutrients seep through the paper). The paper is removed and is dipped in sodium hydroxide to break open the colonies and denature the DNA in the cultures. The denatured (single-stranded) DNA becomes attached to the paper. The sodium hydroxide is neutralized, and radioactive complementary DNA (i.e., complementary to messenger RNA from the gene of interest) is added. Complementary base pairs form between the complementary DNA and paper-attached DNA if the gene being sought is present in the bacterial colony (see Figure 4-10). The paper is washed and placed next to X-ray film (a). If a particular bacterial colony contains the gene of interest [* in (a)], its DNA will have hybridized to the radioactive probe, and the X-ray film will be exposed above the colony (b). The pattern of exposed spots on the film (c) is used to identify the cultures containing cloned genes by comparison with the original distribution of colonies on filter paper (d).*

DNA, it exposes the film, producing a dark spot (Figure AIII-4*b*). These dark spots correspond to bacterial colonies containing the gene of interest. This searching procedure makes it possible to obtain a pure culture of *E. coli* in which each cell contains a plasmid onto which the gene of interest has been cloned.

GLOSSARY

Boldface words are listed as separate entries in the glossary.

adenine: one of the **bases** that is part of one of the **nucleotide** links in a **DNA** or **RNA** chain. It is usually abbreviated by the letter A.

agar: a gelatinlike substance obtained from seaweed. Used in **petri dishes** to culture **bacteria**, agar allows microbiologists to obtain bacterial **colonies**.

agar plate: a **petri dish** containing solid **agar**.

amino acid: a building block of **protein**. The 20 different amino acids have a common structure shown below. The letter R represents chemical side chains, which are different for each amino acid.

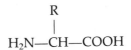

amino acids	
alanine (ala)	leucine (leu)
arginine (arg)	lysine (lys)
asparagine (asn)	methionine (met)
aspartic acid (asp)	phenylalanine (phe)
cysteine (cys)	proline (pro)
glutamine (gln)	serine (ser)
glutamic acid (glu)	threonine (thr)
glycine (gly)	tryptophan (trp)
histidine (his)	tyrosine (tyr)
isoleucine (ile)	valine (val)

aminoacyl-tRNA synthetases: members of a class of **enzymes** that link specific **amino acids** with specific **transfer RNA molecules**. One synthetase recognizes one particular type of transfer RNA and one particular type of amino acid.

ampicillin: an **antibiotic** related to **penicillin**. Both these drugs kill **bacteria** by preventing **cell wall** synthesis.

antibodies: **proteins** that recognize and bind to **antigens**. Antibodies are an important component of the immune system.

antibiotic: a substance that kills an **organism**. In the present context an antibiotic is a drug that kills **bacteria**. Common examples are streptomycin, erythromycin, **penicillin**, and **tetracycline**.

antibiotic-resistance gene: a **gene** that codes for a **protein**, which then allows a **bacterium** to live in the presence of a drug that normally would kill it. **Plasmids** often contain such genes.

anticodon: a particular three-**nucleotide** region in **transfer RNA** that is **complementary** to a specific three-nucleotide **codon** in **messenger RNA**. Alignment of codons and anticodons is the basis for ordering **amino acids** in a **protein** chain.

antigen: a chemical or microorganism that is recognized by, and attaches to, an antibody.

assay: a way to detect something.

atom: a particle composed of a nucleus (protons and neutrons) and electrons. Common atoms are carbon, oxygen, nitrogen, and hydrogen. These atoms differ from one another by having different numbers of protons, neutrons, and electrons. Groups of atoms bonded together produce **molecules**.

ATP: adenosine triphosphate. ATP has high energy bonds that are easily broken by **enzymes** to release the energy needed to drive some of the cellular **chemical reactions**.

attenuation: a type of gene control in which a signal present in the **messenger RNA** can stop **RNA polymerase** from making the messenger RNA longer. The signal is activated by high concentrations of the product of the reactions controlled by the genes being transcribed into the messenger RNA.

B lymphocyte: a type of **cell** in mammals that produces **antibodies**.

bacterial culture: a batch of bacterial cells grown either on solid agar on in a brothlike solution.

bacteriophage: a **virus** that attacks **bacteria**; also called a **phage**.

bacterium (pl: **bacteria**): A one-celled organism lacking a nucleus, mitochondria, and chloroplasts. Although many biochemical properties of bacteria differ from those of higher **organisms**, the basic features of **chemical reactions** are very similar in bacteria and man.

base: (1) A flat ring structure, containing nitrogen, carbon, oxygen, and hydrogen that forms part of one of the **nucleotide** links of a **nucleic acid** chain. The bases are adenine, thymine, guanine, cytosine, and uracil commonly abbreviated A, T, G, C, and U. (2) A hydrogen ion acceptor, like **sodium hydroxide** (lye).

base pair: two **bases**, one in each strand of a double-stranded **nucleic acid molecule**, which are attracted to each other by weak chemical interactions. Only certain pairs form: A-T, G-C, and A-U. See Figure 3-3.

broth: a liquid culture medium used to grow **bacteria**. One common type contains yeast extract, beef extract, table salt, and water.

cell: the smallest unit of living matter potentially capable of self-perpetuation; an organized set of **chemical reactions** capable of self-production. A cell is bounded by a membrane that separates the inside of the cell from the outer environment. Cells contain **DNA**, where information is stored, **ribosomes**, where **proteins** are made, and mechanisms for converting energy from one form to another.

cell extract or lysate: a mixture of cellular components obtained by mechanically or enzymatically breaking **cells**. The cell extract is the starting material from which biochemists obtain **enzymes**, **RNA**, and **DNA**.

cell wall: a thick, rigid structure surrounding certain types of **cells**, especially bacterial and plant cells. Cell walls are often composed of complex **sugars**.

centrifuge: a machine used to create gravitylike forces. Centri-fuges can create forces hundreds of thousands times that of gravity, making it possible to quickly separate **molecules** on the basis of size and shape. Merry-go-rounds and the spin cycle mechanisms of automatic clothing washing machines are examples of centrifuges.

centrifuge rotor: the part of a centrifuge that holds test tubes and rotates at high speed.

chemical reaction: a rearrangement of **atoms** to produce a set of **molecules** that are different from the starting molecules.

chimeric DNA molecule: a **DNA molecule** containing two or more regions of different origin. (e.g., **plasmid** DNA spliced to a fragment of human DNA).

chromosome: a subcellular structure containing a long, dis-crete piece of **DNA** plus the **proteins** that organize and compact the DNA.

clone: (1) noun: a group of identical **cells**, all derived from a single ancestor. (2) verb. to undergo the process of creating a group of identical cells or identical DNA molecules derived from a single ancestor.

cloning vehicles: small **plasmid**, **phage**, or animal **virus** DNA molecules used to transfer a DNA fragment from a test tube into a living cell. Cloning vehicles are capable of multiplying inside living cells. Thus, if a cloning vehicle transfers a specific fragment of DNA into a cell that is also multiplying, all the progeny of that cell will contain identical copies of the vehicle and the transferred **DNA** fragment.

codon: three **nucleotides** whose precise order corresponds to one of the 20 **amino acids**. In addition, special codons that do not code for any amino acid act as stop signals. In some cases several different codons code for the same amino acid.

cointegrate: a type of **DNA molecule** thought to be an inter-mediate in **transposition**. Cointegrates contain the donor DNA, the recipient DNA, and two copies of the **transposon**.

colony: a visible cluster of **bacteria** on a solid surface. All members of the colony arose from a single parental cell, and the colony is considered to be a **clone**. All members are identical. A colony generally contains millions of individual **cells**.

complementary: describing two objects having shapes that allow them to fit together very closely: plugs and sockets, locks and keys, A's and T's or U's, G's and C's.

complementary base-pairing rule: Only certain nucleotides can align opposite each other in the two strands of DNA: G pairs with C; A pairs with T (or U in RNA).

complementary DNA (cDNA): DNA synthesized from **RNA** in test tubes using an **enzyme** called **reverse transcriptase**. The DNA sequence is thus **complementary** to that of the RNA. Complementary DNA is usually made with **radioactive nucleotides** and is used as a **hybridization** probe to detect specific RNA or DNA **molecules**.

conjugation: a **plasmid**-mediated process of pairing between two **bacteria** with the transfer of genetic information (**DNA**) from the plasmid-containing bacterium to the plasmid-lacking bacterium.

constant region: a region of an **antibody molecule** that is identical from one antibody molecule to another in a given class of antibody.

cytosine: one of the **bases** that forms a part of **DNA** or **RNA**. It is usually abbreviated with the letter C.

denature: to unfold, to become inactive. In reference to **DNA**, denaturation means conversion of double-stranded DNA into single-stranded DNA. In reference to **proteins**, denaturation means unfolding of the protein.

density gradient: a solution in which the density gradually increases from top to bottom. Density gradients are commonly used to help separate large **molecules** in a **centrifuge**.

dissolve: to disperse a solid substance in a liquid.

DNA: deoxyribonucleic acid. DNA is a long, thin, chainlike **molecule** that is usually found as two **complementary** chains and is often hundreds to thousands of times longer than the **cell** in which it resides. The links or subunits of DNA are the four **nucleotides** called adenylate, cytidylate, thymidylate, and guanylate. The precise arrangement of these four subunits, repeated many times, is used to store all the information necessary for life.

DNA ligase: the **enzyme** that joins two separate **DNA molecules** together end to end.

DNA polymerase: the **enzyme** complex that makes new **DNA** using the information contained in old DNA.

DNA replication: the process of making **DNA**. DNA is always made from preexisting information in DNA (or, in special cases, from **RNA**). DNA replication involves a number of different **enzymes**.

E. coli: Escherichia coli. These **bacteria** are commonly found in the digestive tracts of many mammals including humans.

egg: **germ cell** produced by a female.

electron micrograph: a photograph taken using an electron microscope, an instrument similar to a light microscope but one which uses a beam of electrons to expose the film rather than a beam of light. Because the effective wavelength of electrons is much shorter than that of light, objects that are measured in millionths of a centimeter can be seen using an electron microscope.

embryo: a plant or animal in an early stage of development, generally still contained in the seed, egg, or uterus.

encode: contain a **nucleotide sequence** specifying that one or more specific **amino acids** are incorporated into a **protein**.

enzyme: a **protein molecule** specialized to catalyze (accelerate) a biological **chemical reaction**. Enzymes are generally named so the last three letters are *-ase*.

equilibrium: the absence of any *net* movement one way or another.

F: the abbreviation for **fertility factor**; a type of transmissible **plasmid**.

Fertility factor: F, a type of transmissible **plasmid** found in *E. coli*.

fetus (adj **fetal**): an **embryo** in a late stage of development, but still in the uterus.

fission: a type of cell division in which a parental cell divides in half to form two daughter cells.

frameshift: displacement of the **nucleotide** reading frame in **DNA** or **RNA**. Frameshifts are generally caused by addition or deletion of one or more nucleotides. Since the nucleotides are read in units of three, addition or deletion of three nucleotides will have no effect on the reading frame.

gel electrophoresis: a method for separating **molecules** based on their size and electrical charge. Molecules are forced to run through a gel (e.g., gelatin) by placing them in an electric field. The speed at which they move depends on their size and charge. See Figure 5-4.

gene: a small section of **DNA** that contains information for construction of one **protein molecule** or in special cases for construction of **transfer RNA** or **ribosomal RNA**.

gene cloning: a way to use microorganisms to produce millions of identical copies of a specific region of **DNA**.

gene expression: the process of transferring information, via **messenger RNA**, from a specific region of **DNA**, a **gene**, to **ribosomes** where a specific **protein** is made.

genetic engineering: the manipulation of the information content of an **organism** to alter the characteristics of that organism. Genetic engineering may use simple methods like selective breeding or complicated ones like **gene cloning**.

germ cells: particular type of **cells** responsible for creating the next generation; also called gametes. In most higher organisms body cells contain two sets of **chromosomes**; germ cells contain only one set. Thus when two germ cells join together, the resulting cell (zygote) has two sets of chromosomes. This cell then produces new body cells.

globin: one of the **protein** chains that comprise **hemoglobin**.

guanine: one of the **bases** that forms a part of **DNA** or **RNA**. It is usually abbreviated with the letter G.

heavy chain: the larger of the two types of **protein** chain that comprise an **antibody**.

hemoglobin: the blood **protein** responsible for transporting oxygen to the tissues.

heteroduplex mapping: the process of locating regions in one DNA molecule that are **homologous** to those in another. Two DNAs are **denatured** and allowed to form **hybrids**. Homologous regions become double stranded while nonhomologous regions remain single stranded. The hybrids are examined by electron microscopy, and the lengths of the double-stranded and single-stranded regions are measured.

homologous: corresponding or similar in position; describing regions of **DNA molecules** that have the same **nucleotide sequence**. Since DNA has two **complementary** strands, **complementary base-pairing** can occur between homologous regions in two different DNA molecules.

host: an **organism** that provides the life support system for another **organism, virus,** or **plasmid.** *E. coli* is a host for certain plasmids that exist inside the **bacterium**, and we are the host for *E. coli*, for it often lives inside us.

hybrid: a double-stranded **nucleic acid** in which the two **complementary** strands differ in origin.

hydrogen bond: a weak attractive force in which a hydrogen **atom** of one **molecule** is drawn toward another molecule.

hypervariable region: a short section of **amino acids** in an **antibody molecule** that frequently exhibits differences in amino acid sequence from one antibody molecule to another.

infectious: capable of invading a **host**.

insulin: a **protein** involved in the control of **sugar metabolism** in mammals. Insulin is made by cells of the pancreas.

interferon: a **protein** made in the body that helps fight **virus** infections.

intron: a noncoding section of a **gene** that is removed from RNA before translation in cells from higher organisms. Bacterial messenger RNA does not contain introns.

lamda (λ): the name of a particular **bacteriophage** used extensively in **gene cloning**.

leader: a section of film that precedes the first scene; a section of **RNA** that precedes the region of the **gene** coding for **protein**.

light chain: the smaller of the two types of **protein** chain that comprise an **antibody**.

lye: sodium hydroxide (NaOH); caustic soda. Dilute lye solutions will cause **DNA** strands to separate, thus producing single-stranded DNA.

lysogen: a **bacterium** harboring a **phage** that does not kill the bacterium and also protects it from further infection by other related phages. A phage called **lambda** (λ) often causes *E. coli* cells to become lysogens.

lysozyme: an **enzyme** that breaks down bacterial **cell walls**. Lysozyme can be obtained from egg white or tears.

lytic infection: a type of **viral** infection in which the cell being attacked is killed and then disintegrates (lyses). In contrast, a lysogenic infection does not lead to cell death.

messenger RNA (mRNA): **RNA** used to transmit information from a **gene** on **DNA** to a **ribosome** where the information is used to make **protein**.

metabolism: a collective term for all the **chemical reactions** involved in life. For example, **sugar** metabolism includes the reactions that occur in the body during the production, use, and breakdown of sugars.

milligram: 0.001 gram (28 grams = 1 ounce).

millimeter: 0.001 meter (1 meter = about 39 inches).

missense: a type of mutation in which a nucleotide change causes an incorrect amino acid to be incorporated into a protein.

mitochondrion (pl. **mitochondria**): a subcellular structure specialized to convert chemical energy from one form to another.

molecule: a group of **atoms** tightly joined together. The arrangement of atoms is very specific for a given molecule, and this arrangement gives each molecule specific chemical and physical properties. The oxygen molecule we breathe is two oxygen atoms bonded together. Paper is largely cellulose molecules, which are giant molecules containing carbon, oxygen, and hydrogen.

mutagen: an agent that increases the rate of **mutations** by causing changes in the **nucleotide sequences** of DNA.

mutant: an **organism** whose **DNA** has been changed relative to the **DNA** of the dominant members of the population.

mutations: errors in **DNA**, often occurring during **DNA replication**, that cause incorrect **amino acids** to be inserted into **proteins**.

nuclease: a general term for an **enzyme** that cuts **DNA** or **RNA**.

nucleic acid: DNA, RNA, or DNA:RNA hybrid.

nucleoid: a compact, **DNA**-containing structure found in **bacteria**; the bacterial equivalent of a **chromosome**.

nucleotide: one of the building blocks of **nucleic acids**. A nucleotide is composed of three parts: a **base**, a **sugar**, and a **phosphate**. The sugar and the phosphate form the backbone of the nucleic acid, while the bases lie flat like steps of a staircase. **DNA** is composed of four different kinds of nucleotide represented by the letters A, T, G, and C. See also **sequence**.

nucleotide pair: two **nucleotides**, one in each strand of a double-stranded **nucleic acid** molecule, which are attracted to each other by weak chemical interactions between the **bases**. Only certain pairs form: AT, GC, and AU.

nucleus: (1) the core of an **atom** consisting of protons and neutrons; (2) a distinct subcellular structure containing **chromosomes**.

operator: a region on **DNA** capable of interacting with a **repressor**, thereby controlling the functioning of an adjacent **gene**.

operon: a series of **genes** transcribed into a single **RNA** molecule. Operons allow coordinated control of a number of genes whose products have related functions.

organism: one or more **cells** organized in such a way that the unit is capable of reproduction.

origin of replication: a special **nucleotide sequence** that serves as a start signal for **DNA replication**.

pathogen: a disease-causing agent (e.g., **viruses** that cause polio, mumps, and measles; **bacteria** that cause cholera and leprosy).

penicillin: an **antibiotic** that kills *E. coli* by blocking formation of new **cell walls**. Penicillin is produced by a mold.

peptide bond: the type of chemical bond that links two adjacent **amino acids** together in a **protein** chain.

phage: Bacteriophage.

phage plaques: clear zones, created by **bacteriophages** killing **bacteria**, in a lawn of bacteria on an **agar plate**. By counting the number of plaques, it is possible to determine the number of **phages** present before pouring a mixture of bacteria and phages on the agar plate.

phenol: an oily organic chemical used to separate **DNA** from **proteins**. Phenol is added to a mixture of DNA and proteins in water and the mixture is vigorously shaken. Proteins tend to move into the phenol; DNA stays in the water phase. The mixture is then briefly centrifuged. Phenol is more dense than water, so it forms a layer under the water. The water layer, containing the DNA, is removed with a pipette, leaving proteins behind in the phenol.

phosphate: a chemical unit in which four oxygen **atoms** are joined to one phosphorus atom. The backbones of **DNA** and **RNA** are alternating phosphate and **sugar** units.

pilus (pl. **pili**): a long, hairlike structure produced on the surface of a **bacterium** containing a certain type of **plasmid**. Plasmid **genes** code for the **proteins** that comprise the pilus.

pipette: a long, thin glass tube used for measuring volumes of liquids. A pipette can be used like an eyedropper to add or remove liquids from test tubes.

plasmids: small circular **DNA molecules** found inside **bacterial cells**. Plasmids reproduce every time the bacterial cell reproduces. Once infected, the bacteria will always contain a plasmid.

point mutation: change of only one **nucleotide** pair in a **DNA** molecule.

poly A or **poly T tail:** A long stretch (more than 20) of pure A's or T's, respectively, at the end of a DNA or RNA strand.

precipitate: **molecules** that are clumped together so that they fail to pass through a filter. Precipitates are large aggregates and settle out of solution rapidly, much like silt out of river water.

primer: a piece of **DNA** or **RNA** that provides an end to which **DNA polymerase** can add **nucleotides.**

probe: a **radioactive RNA** or **DNA** molecule used to identify **bacterial colonies** that contain **cloned genes**.

product: the new **molecules** produced by a **chemical reaction**.

progeny: offspring.

promoter: a short **nucleotide sequence** on DNA where **RNA polymerase** binds and begins **transcription**.

protease: an enzymatic **protein** that breaks down other proteins.

protein: a class of long, chainlike molecules often containing hundreds of links called **amino acids**. There are 20 different amino acids used to make proteins. The thousands of different proteins serve many functions in the **cell**. As **enzymes** they control the rate of **chemical reactions**, and as structural elements they provide the cell with its shape. Members of this group of molecules are also involved in cell movement and in the formation of cell walls, membranes, and protective shells. Some proteins also help package the long **DNA** molecules into **chromosomes**.

protein synthesis: see **translation**.

pseudogene: a nonfunctional **gene** that closely resembles a functional one.

purify: to separate one type of **molecule** away from other types.

radioactive: describing a substance (a **molecule** in the context of this book) containing an unstable element that spontaneously emits a high energy particle or radiation. The emission is detectable by photograhic film, by Geiger counter, and by other sophisticated instruments. Gene cloners generally use radioactive hydrogen, carbon, or phosphorus, all of which are commercially available. Radioactive uranium and plutonium are used in nuclear reactors.

radioactive tracer: a **radioactive atom** that is incorporated into a specific **molecule** such that the molecule can be detected and measured by the presence of the radioactivity. It is much easier to measure small amounts of radioactivity than small amounts of particular chemicals.

recognition site: a short series of **nucleotides** specifically recognized by a **protein**, usually leading to the binding of that protein to the **DNA** at or near the point of the recognition sequence. Once the protein has bound to the DNA, it may cut, modify, or cover the DNA, depending on the function of the protein.

recombinant DNA molecule: a chimeric DNA molecule formed by cutting and splicing technologies.

recombination: the breaking and rejoining of **DNA** strands to produce new combinations of DNA molecules. Recombination is a natural process that generates genetic diversity. Specific **proteins** are involved in recombination.

replica plating: the process in which **colonies** of **bacteria** are transferred from one **agar plate** to another without changing their relative orientation.

replication fork: the point at which the two parental **DNA** strands separate during **DNA replication**.

repression: a method of preventing **gene expression** in which a **protein** molecule (**repressor**) binds to the **DNA** near where **RNA polymerase** ordinarily would bind.

repressor: a **protein** molecule that is capable of preventing **transcription** of a **gene** by binding to **DNA** in or near the gene.

restriction endonucleases: **enzymes** that cut **DNA** at specific **nucleotide** sequences. The function of this class of enzyme inside **cells** is to protect the cells against invasion by foreign DNA.

restriction mapping: a procedure that uses **restriction endonucleases** to produce specific cuts in **DNA**. The positions of the cuts can be measured and oriented relative to each other to form a crude map.

reverse transcriptase: an **enzyme** purified from tumor **viruses** that makes **DNA** from **RNA**.

ribosomes: large ball-like structures that act as workbenches where **proteins** are made. A bacterial ribosome consists of two balls, a small one called 30*S* and a larger one called 50*S*. Ribosomes are composed of special **RNA molecules** (ribosomal RNA) and about 50 specific proteins (ribosomal proteins).

RNA: ribonucleic acid. RNA is a long, thin, chainlike **molecule** usually found as a single chain. The links, or subunits of RNA, are the four **nucleotides** abbreviated as A, U, G, and C. Cells contain a number of different kinds of RNA molecules, which play different roles. The most common RNAs are **messenger RNA, transfer RNA,** and ribosomal RNA. Some **viruses** use RNA as their genetic material.

RNA:DNA hybrid: a double-stranded molecule composed of one strand of **RNA** and one of **DNA**. The **nucleotide sequences** in the DNA and RNA are **complementary**.

RNA polymerase: the **enzyme** complex responsible for making **RNA** from **DNA**. RNA polymerase binds at specific **nucleotide sequences** (**promoters**) in front of **genes** in DNA. It then moves through a gene and makes an RNA molecule that contains the information contained in the gene. RNA polymerase makes RNA at a rate of 65 nucleotides per second.

RNA tumor virus: a type of **RNA**-containing **virus** that produces tumors in animals or converts normal cells in culture into tumor cells.

sequence: the order of. In reference to **DNA** or **RNA**, sequence means the order of **nucleotides**.

sodium hydroxide (NaOH): lye, caustic soda. A chemical that separates DNA strands.

sperm: germ cell produced by a male.

sterile: without life, generally referring to an instrument or a solution that has been heated to kill any **organisms** that may have been on or in it. Wire is sterilized by heating in a flame until it is red hot. Culture medium (e.g., **broth**) is sterilized by heating in a pressure cooker.

sticky ends: specific termini (ends) of double-stranded **DNA** in which one of the strands sticks out farther than the other and the protruding strand is **complementary** to the protruding strand at the other end of the **DNA** molecule or at the end of another DNA molecule.

submicroscopic: not visible when examined with a light microscope.

substrate: the **molecules** on which an **enzyme** acts.

subunit: one of the pieces that forms a part of a multicomponent structure, such as a link in a chain or a brick in a wall.

sugar: a class of **molecule** containing particular combinations of carbon, hydrogen, and oxygen. The sugars in **DNA** and **RNA** are five-carbon sugars called deoxyribose and ribose, respectively. Glucose is a sugar containing six atoms of carbon per molecule.

tetracycline: an **antibiotic** that kills **bacteria** by blocking **protein** synthesis.

thymine: one of the **bases** that forms part of **DNA**. It is not found in **RNA**. It is usually abbreviated by the letter T.

toxin: a substance, often a **protein** in the context of this book, that causes damage to the **cells** of an **organism**.

transcription: the process of converting information in **DNA** into information in **RNA**. Transcription involves making an RNA molecule from the information encoded in the DNA. **RNA polymerase** is the **enzyme** complex that executes this conversion of information.

transfer RNAs (tRNAs): small RNA molecules (about 80 **nucleotides** long) that serve as adapters to position **amino acids** in the correct order during **protein** synthesis. The ordering by tRNA uses information in **messenger RNA** and occurs before the amino acids are linked together.

transformation: the process whereby a **bacterial cell** takes up free **DNA** such that information in the free DNA becomes a permanent part of the bacterial cell.

translation: the process of converting the information in **messenger RNA** into **protein**. Also called protein synthesis.

transposase: a **protein** encoded by a **gene** in a **transposon** and required for **transposition**.

transposition: the process whereby one region of **DNA** moves to another. Transposition generally involves duplication of the region that moves.

transposon: a short section of **DNA** capable of moving to another DNA molecule or to another region of the same DNA molecule.

tryptophan: one of the 20 **amino acids** found in **proteins**.

ultraviolet light: a type of light that has very high energy and is invisible; black light. **Nucleic acids** absorb ultraviolet light, and instruments are available that measure the amount of absorption. The amount of absorption depends on the amount of nucleic acid present. Thus, by measuring the amount of absorption, it is possible to measure the amount of nucleic acid present.

uracil: one of the **bases** that forms part of **RNA**. It is generally not found in **DNA**. It is usually abbreviated with the letter U.

variable region: a region of an **antibody molecule** that differs in **amino acid sequences** from one antibody to another. Variable regions are thought to exist where *antigens* bind to antibodies.

virus particles: a class of infectious agents usually composed of **DNA** or **RNA** surrounded by a protective **protein** coat.

walking along DNA: a **gene cloning** procedure used to clone regions of **DNA** adjacent to ones already cloned.

yeast: one-celled **organism** containing a true **nucleus** and **mito-chondria**. Many biochemical properties of yeast are similar to those of higher organisms, and yeast probably is more closely related to mammals than to **bacteria**.

Index

201

FLOW OF GENETIC INFORMATION
replication, transcription, and translation

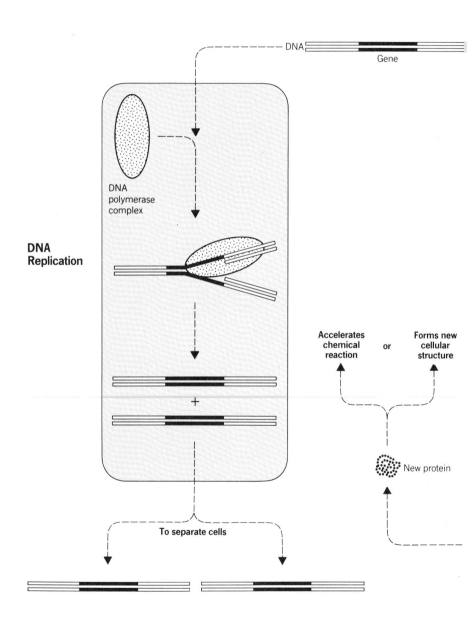

DNA

Gene

DNA polymerase complex

DNA Replication

Accelerates chemical reaction

or

Forms new cellular structure

New protein

+

To separate cells